MEMOIRS OF THE MUSEUM OF ANTHROPOLOGY
UNIVERSITY OF MICHIGAN
NUMBER 17

**Studies in Latin American
Ethnohistory & Archaeology**

Joyce Marcus, General Editor

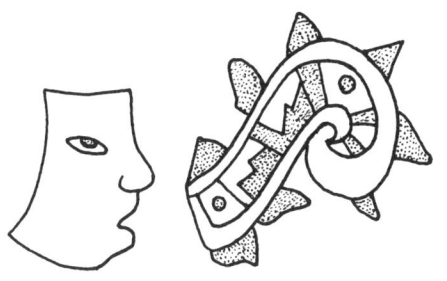

Volume 2

Irrigation & the Cuicatec Ecosystem: A Study of Agriculture & Civilization in North Central Oaxaca

by

Joseph W. Hopkins III

ANN ARBOR
1984

This series is partially supported by a grant-in-aid No. 4453 from the Wenner-Gren Foundation for Anthropological Research, whose Director of Research, Lita Osmundsen, offered both encouragement and help during the preparation of the grant proposal. Generous funds were also supplied by the Museum of Anthropology, University of Michigan, through the efforts of former Director Richard I. Ford.

© Regents of the University of Michigan
The Museum of Anthropology
All Rights Reserved

Printed in the
United States of America

ISBN 0-915703-00-9

To my father
"Curse, bless, me now with your fierce tears . . ."

Introduction to Volume 2

by
Joyce Marcus

This *Memoir* constitutes a companion piece to Volume 1, "A Fuego y Sangre: Early Zapotec Imperialism in the Cuicatlán Cañada, Oaxaca," by Elsa M. Redmond (1983). These two volumes, along with that by Charles S. Spencer (1982), represent significant increases in our knowledge of one riverine canyon—the Cañada de Cuicatlán—which has a strategic location in the southern Mexican highlands, linking as it does the Tehuacán Valley to the Valley of Oaxaca. In one decade three major projects—those directed by Hopkins, Redmond, and Spencer—took this canyon as the central focus of study.

One of the interesting aspects of these three projects is that they were carried out in an area most Mesoamerican archaeologists had left fallow for decades, primarily because so little was known about it. The boon of such research is the special perspective it affords us—the opportunity to view the operations of the pre-Columbian state from its periphery or "frontier zone" which at some times may have been quite autonomous, other times tributary, and still other times subjugated completely. The dynamic expansion and contraction of the state that takes place along its borders can be studied in just such a place as the Cañada de Cuicatlán.

In contrast to the Tehuacán Valley and the Valley of Oaxaca whose valley floors lie ca. 1500–1700 m in elevation, the Cuicatlán Cañada valley floor varies from ca. 500–800 m. These latter elevations fall squarely within *tierra caliente*, rather than the *tierra templada* and *tierra fría* zones that surround it. The sixteenth-century Relación de Cuicatlán states that this area is noted for growing the best tropical fruits in New Spain (Gallego 1580:187). As Redmond (1983) and Spencer (1982) have argued (Redmond and Spencer 1983), this *tierra caliente* canyon's ability to grow desired fruits and other tropical products may have been one of the factors that led to this area being the object of conquest and tributary demands at various times during its checkered past.

While Redmond focuses on the subjugation of this canyon during the Terminal Formative (ca. 200 B.C.–A.D. 200) by the Zapotec capital Monte Albán in the Valley of Oaxaca, Hopkins (in this volume) devotes his attention to the relationship between the expansionistic Aztec empire and the Cuicatlán Cañada during the Late Postclassic period (ca. A.D. 1450–1519). Both Redmond and Hopkins attempt to understand local Cañada de Cuicatlán events as part of long-term regional processes, which reveal a multi-ethnic and a multi-valley character. Hopkins recognizes the Cuicatec ecosystem of highland and lowland towns as but one unit of study within the larger system he designates the "macrosystem".

In the sixteenth century, the Cañada de Cuicatlán was quite similar in political organization to the Mixteca Alta. Both areas were characterized by a number of separate and autonomous polities called *cacicazgos*. The Cuicatec *cacicazgos* were linked by royal marriage alliances, *cacique* and *cacica* sometimes linked by language and blood as part of an endogamous class, or linked to Mixteca Alta nobility. Some of the Cuicatec towns were apparently conquered by the Mixtec, and they, thereafter, owed them tribute. Other communities in the Cuicatec area were multi-ethnic: including Mixtec and Cuicatec, others comprised of Chinantec and Cuicatec, and still others included Mazatec and Cuicatec.

Just as the Zapotec before them, the Aztec made tribute demands on the Cuicatlán Cañada towns. Cuicatec royal marriage alliances linking *cacicazgos* as well as the multi-ethnic composition of various communities in the Cuicatec area produced the mosaic background upon which we must place and evaluate the nature and impact of the Aztec empire. Some Cuicatec towns were administered by Mixtec *cabeceras* (head towns), and following their conquest at the hands of the Aztec, the tributary hierarchy passed the goods up to the newly-imposed top level of the inter-regional hierarchy.

Hopkins emphasizes that we must consider how this Cuicatec ecosystem of highland and lowland towns functioned within the macrosystem of Mesoamerica under the Aztec empire. While this administrative overlay is just one component of the macrosystem, Hopkins provides evidence which suggests that the Aztec (like the Spaniards) left much of the local Cuicatec political and economic structures intact. At the same time, Hopkins is cautious in his reconstruction of the Cuicatec political and economic systems, aware that all our ethnohistorical documents postdate both the Aztec and Spanish Conquests.

What Hopkins presents is a view of Cuicatec *cacicazgos* as part of a nested set of systems with three hierarchical levels—the Mesoamerican macrosystem at the top, the Cuicatec ecosystem of highland and lowland towns in the middle, and the individual Cuicatec *cacicazgos* at the bottom.

Unlike most archaeological/ethnohistorical studies that often restrict their temporal focus to the sixteenth century or period of contact, Hopkins extends his ethnohistorical study into the historic era (seventeenth through nineteenth centuries), and up to the present day. Thus, we know that (1) the population in the Cuicatec area reaches its nadir in the eighteenth century, with the lowlands suffering far greater population losses than the highlands; (2) by the twentieth century the lowlands were completely "Spanish," while the highlands remained "Indian"; (3) the lowlanders continue to practice irrigation agriculture, now with a commercial export bent, while the highlanders mostly practice rainfall subsistence agriculture; and (4) contemporary (and past) lowland towns are on major transportation routes, and are relatively prosperous because of their tropical fruit exports; the contemporary highland towns are still fairly inaccessible, less prosperous, and have fewer external contacts and transactions. This intra-Cuicatec set of comparisons between the highlands and the lowlands from the Spanish Conquest to the present is a welcome feature, an exercise in which more ethnohistorians should engage.

Hopkins evaluates both the Postclassic (Iglesia Vieja phase) remains and the degree of fit between them and the ethnohistorical model he developed in Chapter 2; he then turns his attention to Postclassic agricultural and irrigation systems, a major focus of his research. During this period there is evidence for dense occupation in virtually every locality where irrigation systems were possible. Data collected by Hopkins indicate that (1) irrigation systems were considerably more extensive in the Postclassic period than any in use today, and (2) the Postclassic irrigation canals took off the water higher than at present, in order to make lands which are unreachable today productive in the pre-Hispanic era. Some of the evidence consists of mortar and stone canals and aqueducts reaching lands that are unirrigated today. Where it was possible to date these features, they proved to be Postclassic, and many of those were Late Postclassic. Additionally, in the presently-unoccupied intermediate zone lying between the highlands and the lowlands, Hopkins located important irrigation and terrace systems in the small canyons of the Ríos Chiquitos and the headwaters of the Río Grande.

Postclassic sites frequently occur on the high natural promontories, a pattern often associated with militarism, defense, and protection against the encroachment of enemies. Hopkins (1983) suggests still another possibility—that the Cuicatec did not want to waste any potentially irrigable land. Two kinds of towns comprise the Cuicatec settlement system, highland and lowland, and they apparently enjoyed a complementary or "symbiotic" relationship, exchanging items that could be produced in each of these two very different environmental zones. In the Postclassic period he also sees an increase in population, evidence for occupation in previously unoccupied areas, and a greater reliance on intensive forms of cultivation. When the Cuicatec ecosystem had lower populations, irrigation was a less prominent feature; as towns and population grew, irrigation became a more prominent feature. However, what happened in the Cañada de Cuicatlán (increasing population and a greater reliance on intensive agriculture) was part of a Postclassic highland Mesoamerican trend. As intensification increased in the Cañada de Cuicatlán it took various forms, such as that seen at El Despoblado in which canals and terraces were constructed. The need for more land, and for more than one crop per year, were presumably behind these forms of intensification.

An additional component of the Postclassic Cuicatec system was an intraregional exchange which was articulated with other economic state systems. Eva Hunt (1972:240) has argued that the source of Cuicatec political power resided in the ability of certain individuals to control the exchange networks linking highlands and lowlands as well as land reallocation, water distribution, labor, temple activities, and defensive strategies. Irrigation itself, argues Hopkins, was not the "cause" of social stratification in the Cañada de Cuicatlán, nor the basis of power. Hopkins concludes, much as Hunt did, that the control of local surpluses within the Cuicatec ecosystem, and the channeling of those goods into larger systems (e.g. the Zapotec or Aztec "states"), was the basis for power and the maintenance for social stratification.

The sixteenth-century peoples in the state of Oaxaca present an exciting spectrum of sociopolitical complexity (Marcus and Flannery 1983:222) from the highly stratified Mixtec to the lowly Chontal whom the Spaniards considered "as brutish as the wild deer in the mountains" (Diez de Miranda 1579:26). The Cuicatec *cacicazgos*, as Hopkins has reconstructed them, would fall somewhere in between, sometimes more closely mimicking the stratified Mixtec system and at other times behaving more like simpler "community kingdoms."

Bibliography

Diez de Miranda, Gutierre
    1579   Relación de Xuchitepec. In *Papeles de Nueva España: Segunda Serie, Geografía y Estadística*, Vol. 4, Francisco del Paso y Troncoso, ed., pp. 24–28. Madrid: Sucesores de Rivadeneyra (1905).

Gallego, Juan
    1580   Relación de Cuicatlán. In *Papeles de Nueva España: Segunda Serie, Geografía y Estadística*, Vol. 4, Francisco del Paso y Troncoso, ed., pp. 183–89. Madrid: Sucesores de Rivadeneyra (1905).

Hopkins, Joseph W. III
    1983   The Tomellín Cañada and the Postclassic Cuicatec. In *The Cloud People: Divergent Evolution of the Zapotec and Mixtec Civilizations*, Kent V. Flannery and Joyce Marcus, eds., pp. 266–70. New York: Academic Press.

Hunt, Eva V.
    1972   Irrigation and the Socio-Political Organization of the Cuicatec Cacicazgos. In *The Preshistory of the Tehuacán Valley*, Vol. 4: *Chronology and Irrigation*, Frederick Johnson, ed., 162–248. Austin: University of Texas Press.

Marcus, Joyce, and Kent V. Flannery
    1983   An Introduction to the Late Postclassic. In *The Cloud People: Divergent Evolution of the Zapotec and Mixtec Civilizations*, Kent V. Flannery and Joyce Marcus, eds., pp. 217–26. New York: Academic Press.

Redmond, Elsa M.
    1983   A Fuego y Sangre: Early Zapotec Imperialism in the Cuicatlán Cañada, Oaxaca. In *Studies in Latin American Ethnohistory & Archaeology*, Vol. 1, Joyce Marcus, ed. *Memoirs of the University of Michigan Museum of Anthropology* 16.

Redmond, Elsa M., and Charles S. Spencer
    1983   The Cuicatlán Cañada and The Period II Frontier of the Zapotec State. In *The Cloud People: Divergent Evolution of the Zapotec and Mixtec Civilizations*, Kent V. Flannery and Joyce Marcus, eds., pp. 117–20. New York: Academic Press.

Spencer, Charles S.
    1982   *The Cuicatlán Cañada and Monte Albán: A Study of Primary State Formation*. New York: Academic Press.

Contents

Illustrations xi
Tables xii
Acknowledgments xiii

1. **Introduction** 1
   Cuicatlán, the Cañada, and the Cuicatecs 1
   The Cuicatec Region 2

2. **The Cuicatec Ecosystem: An Historical Reconstruction** 9
   Historical Sources 9
      Codices 9
      Spanish Sixteenth-Century Sources 9
      Later Sources 11
      Recent Works 12
   Definition of the Cuicatec Region 12
   The Cuicatec Ecosystem 13
      The Cañada Towns 18
      Integration of the Cuicatec Region 22
      Social and Political Structure 25
      The Cuicatec in the Aztec Empire 31

3. **The Spanish Conquest: The Cuicatec in Culture Contact** 37
   Culture Contact and Schismogenesis 37
   The Spanish Conquest of the Cuicatec Region 38
   The Impact of the Spanish Conquest 39
      Post-Hispanic Population Change and the Cuicatec Ecosystem ... 43
      Estimating Population for the Cuicatec Ecosystem 43
   Spanish Introductions to Mesoamerica 59
      Livestock in the Cuicatec Region in the Sixteenth Century 60
      Introduced Plants 62
      Summary: The Impact of the Spanish Conquest 63

4. **The Cuicatec Ecosystem from the Conquest to the Present** 65
   The Seventeenth Century: The Introduction of Sugar Cane 65
   The Eighteenth Century: The Spaniards Establish a Foothold 66
   The Nineteenth Century: Roads, Railroads, and Haciendas 69
   The Cuicatec Ecosystem in the Twentieth Century 71
   Analysis: Processes in the History of the Cuicatec Ecosystem 79
      Interaction of Factors 84

5. Archaeological Evidence: The Cuicatec Ecosystem from its Origin	87
The Preclassic Period	89
Perdido Phase	89
Lomas Phase	89
The Classic Period Trujano Phase	89
The Postclassic Iglesia Vieja Phase	89
Ancient Irrigation Remnants	94
Population, Irrigation Systems, and Settlements in the Cuicatlán Cañada	118
Population and Intensive Agriculture in Mesoamerica	120
Cuicatlán and Mesoamerica	126
Appendix. Archaeological Fieldwork	131
Survey of Cuicatlán, Oaxaca	131
Other Cuicatec Sites	133
El Sabino	133
La Unión (Site 28)	134
Atlatlauca	134
Cotahuixtla	134
El Despoblado and the Río de las Trancas Canals (Site 32)	134
La Coyotera (Site 29)	134
Ojo de Pajarito	134
Santa Cruz	134
Dominguillo	135
Excavations	135
Site 5	135
Site 16	135
Artifacts	136
Bibliography	139
Resumen en Español	145

Illustrations

1. Travel routes through the Cañada 4
2. Population of the Cuicatec region from Spanish contact to the present 57
3. Indians from Reyes Papalo descending to Cuicatlán with *vigas, tablas,* and nets full of charcoal 74
4. Brush dam construction for tomas de aguas 76
5. Land use in Cuicatlán 78
6. *Canoas* 79
7. Archaeological sites in the Cuicatec region 86
8. Sites in the vicinity of Cuicatlán 88
9. Ojo de Agua (Site 21) 90
10. Profile A-B, Ojo de Agua (Site 21) 91
11. Site 16, Cuicatlán 93
12. Site 18, Cuicatlán 94
13. Juego de pelota on Iglesia Vieja meseta 95
14. Building levels, Site 16 96
15. Toma de los Chentiles, on the Río Chiquito, Cuicatlán 98
16. Downstream fragments of the Toma de los Chentiles, high on the slope above the Río Chiquito, Cuicatlán 101
17. Cross section, canal of the Toma de los Chentiles, where it crosses Site 5 102
18. Cross section of trench cutting the canal of Toma de los Chentiles, Site 5 103
19. Plan of Site 5, showing excavations 104
20. Profile of Site 5, Cuicatlán 105
21. House, Site 5 106
22. Northeast wall of house, Site 5 108
23. Step in plaster floor of house, Site 5 109
24. Tombs, Site 5 110
25. Abandoned canal fragment on the Río Cacahuatal, between Cuicatlán and Quiotepec 111
26. Mineralized fragments of an earlier canal beside present canal, Santa Cruz, Almoloyas 113
27. Canal fragments on the Río de las Trancas, or Carrizal 115
28. Terraces fed by canal fragments on the north side of Río de las Trancas, below El Despoblado 116
29. Large mound at El Despoblado 117

Tables

1. Identification of Cuicatec Towns...14
2. Cuicatec Population Data...45
3. Population Sample of Cuicatec Towns...58

Acknowledgments

There is no way I can thank all the people who have helped me complete this study. However, I would like to mention a few. Drs. Eva and Bob Hunt suggested the Cuicatec area for my research. They corresponded with me during the fieldwork and gave me much useful advice and criticism. Eva loaned me copies of her own notes from unpublished and published sources on the Cuicatec and served on my thesis committee, guiding my efforts at ethnohistorical analysis. I felt her loss keenly. Bob Adams helped me outline my proposal, both to the Anthropology Department and to the National Science Foundation. His letters while I was in the field gave me a perspective on my work while I was in the midst of it. After I returned to Chicago, he patiently supervised the various versions of the thesis as it slowly came into being. Les Freeman, the third member of my doctoral committee, also helped with valuable criticism and invaluable moral support.

The fieldwork in Cuicatlán was financed by Doctoral Grant no. GS–2274 from the National Science Foundation, as well as a National Defense Loan. The fieldwork was done under a *convenio* with the Departmento de Monumentos Prehispánicos, of the Instituto Nacional de Antropología e Historia of Mexico. I am grateful to the Arquitecto Ignacio Marquina, the director of this department, for his cooperation and patience for the original *convenio* and for granting me several extensions on the period of the *convenio*. The Arqueólogo Eduardo Matos Moctezuma, subjefe of the department, and Gladys Casimires de Bizuela, a member of the staff of the department and a good friend, helped me understand the requirements of the *convenio* and to obtain extensions with a minimum of misplaced effort.

I could not have done the fieldwork without the aid that I received from many archaeologists who were in Oaxaca while I was in the field. In particular, Kent Flannery and his group from the University of Michigan introduced me to Oaxaca and its archaeology. Kent was always willing to look at material I had recovered and give me invaluable advice on fieldwork.

Ronald Spores of Vanderbilt University, Donald Brockington from the University of North Carolina, and John Paddock and his students from the University of the Americas all helped me at various times with the identification and interpretation of my finds from the Cañada.

Rafael Vásquez Cruz, who was the archaeological inspector for the I.N.A.H. in charge of the zone of the Cañada, deserves every praise for his zealous efforts to preserve the archaeological remains of the zone from destruction by nature or man. He was generous enough to share the extensive knowledge of the area which he had acquired in his many years visiting the zone. My survey of the Cañada would have been at least much less efficient and more probably impossible without his services as a guide. Everyone who has known Rafael has been impressed by his zealousness, intelligence, helpfulness, and humility. I cannot thank him enough for his help.

I would like in addition to thank all the people of Cuicatlán, who put up with me for two years and who were friendly and hospitable to me, whether I was walking through their fields, wandering through their hills, or simply sitting and piling potsherds in my house in the town. I remember fondly the two years I spent in Cuicatlán. I would also like to thank all my friends in Mexico City, who always made me welcome when I came to renew permissions, buy equipment, or simply see the big city.

I have already mentioned the help Eva Hunt gave me with materials for the historical analysis. I would like also to thank the knowledgable staff of the Newberry Library. It was a pleasure using the facilities of the library.

Joyce Marcus, General Editor of this series and this volume, has helped make available unpublished information on more recent research in the Cañada. Joyce has made many valuable suggestions which resulted in considerable improvement in my manuscript.

I thank my sister, Susan, for redrawing my illustrations. Finally, I am grateful to my wife Louise for her patience and support during the endless writing and rewriting. I hope the result in some way justifies the contributions all these people have made.

Chapter 1

Introduction

Cuicatlán, the Cañada, and the Cuicatecs

The interrelationship between agricultural systems and the various processes of civilization in Mesoamerica is the topic of this book. An ecological systems model is constructed using ethnohistorical, ethnographic, and archaeological evidence. The workings of the system over time are then analyzed. This integration of different lines of evidence allows us to explain more powerfully the past. It does away with the artificial barrier between pre-Hispanic and post-Hispanic research in Mesoamerica.

For my research I chose one small pre-Hispanic irrigation system. I wanted a small system I could manage on a modest budget and limited time. An irrigation system was preferred because I wanted to examine specific issues raised by Wittfogel (1957, 1972) as well as more general hypotheses presented by Sanders (1956, 1972), Boserup (1965), and Flannery (1968a, 1972). I did not specify that the irrigation system be early or late, as long as it was pre-Hispanic. At the time I went into the field, it was not clear whether irrigation systems were common in the earlier periods. Because I did not know beforehand which period or periods the irrigation system would span, I expected the project to take one of two forms:

> If vestiges of systems from various epochs are found, then the emphasis of the project will be on the development over time of the various systems of agriculture that have left vestiges, and their relations to the development of the population center. If the remains seem to be primarily from one period, the emphasis will be on a careful reconstruction of one agricultural system as it existed, and a careful reconstruction of the population center it supported. [Hopkins 1968b:5-6]

Drs. Eva and Robert Hunt suggested I consider the Cuicatec area, in north-central Oaxaca, Mexico. They had carried out considerable ethnological and ethnohistorical work in this region. Their work was inspired by an interest in the relationship of political structure to the agricultural system, especially the irrigation system. From the ethnohistorical research they had completed, the Hunts provided evidence that irrigation systems existed in the Cuicatec Cañada before the Spanish Conquest. They offered to share their notes which have been of great help to me.

Another benefit was that the Cuicatec area was surrounded by areas in which recent ecologically-oriented work had been, or was being, done. To the north is the Valley of Tehuacán, where Richard S. MacNeish's Tehuacán Archaeological-Botanical Project was carried out. To the west, Ronald Spores had been directing a multidisciplinary archaeological and ethnohistorical investigation in the Nochixtlán Valley. And to the south, in the Valley of Oaxaca there was a long history of archaeological and ethnohistorical research, including such well-known researchers as John Paddock, Alfonso Caso, Ignacio Bernal, and the multidisciplinary research project conducted by Kent Flannery of the University of Michigan. All of this related work made the Cuicatec area ideal for my research.

In order to analyze the Cuicatec ecosystem, I identified four dimensions: three spatial and one temporal. I defined the temporal dimension as that from the first human habitation of the ecosystem up to the present. I defined the spatial dimension of the ecosystem as the region occupied by Cuicatec speakers at the time of the Spanish Conquest. I chose these dimensions because they made my work congruent with the work of the Hunts.

This definition of the ecosystem allows me to talk about the development and processes of culture within a specific area over time. Because this system is defined by physical dimensions, I am not limited to the presence of Cuicatec speakers. When Cuicatec speakers either were replaced by other populations, or assimilated, I can still consider the new population in my study, because of how I have defined my unit of study as a geographical area.

The Cuicatec Region

Chapter 3 explains how the area that was occupied by Cuicatec speakers at the time of the Spanish Conquest was reconstructed. The northern limit of this area was the Cañada of the Río Santo Domingo, where the combined Salado and Río Grande flow out through the Sierra Madre to form one of the major tributaries of the Papaloapan. To the east, the boundary is roughly that between *tierra fría/templada* and *tierra caliente*, on the eastern slope of the Sierra Madre Oriental. The Cuicatecs were above this line, while the Chinantec occupied the lower, *tierra caliente* slope of the Sierra Madre. To the west the boundary is the Cañada of the Río Grande, the floor of which was Cuicatec. The villages in the highlands to the west of the Cañada were Mixtec. To the south, the Cuicatec region bounded that of the Zapotec, just short of the watershed between the Río de las Vueltas and the Valley of Oaxaca. Farther east, around Atlatlauca, the frontier was with the Chinantec. The two main parts of the Cuicatec ecosystem are the Cuicatec part of the Cañada, and the Cuicatec-occupied crest and western slope of the Sierra Madre Oriental which form the eastern boundary of the Cañada. My research was concentrated in the Cañada.

The Cañada is a deep canyon running from the Tehuacán Valley to just north of the watershed of the Valley of Oaxaca. It is the valley of the Río Salado, running south from the Valley of Tehuacán, and the Río Grande and its tributaries, which run north from the back of the watershed of the Valley of Oaxaca. At Quiotepec, the two rivers meet and form the Santo Domingo. The Santo Domingo leaves the Cañada and runs east, cutting sharply through the Sierra Madre Oriental to the state of Veracruz, where, after being joined by several other rivers, it becomes the Papaloapan, which empties into the Gulf of Mexico.

The floor of the Cañada lies at from 500 to 800 m above sea level. To the east, the Sierra Madre Oriental rises to an average height of over 2000 m, with peaks as high as 3300 m. To the west are the somewhat lower, but still considerable, heights of the Sierra Mixteca, which rise to around 2000 m. The floor of the Cañada ranges from 1 to 10 km in width. It extends about 110 km north and south.

The only detailed geological description of the Cañada is that of Tomás Barrera (1946). According to Barrera, the geological history of the Cañada began when the Cañada was formed by a tertiary "phenomenon" in Cretaceous and pre-Cretaceous rock. This faulting is indicated by the fact that the eastern border of the canyon is formed of pre-Cretaceous metamorphic rocks capped by Cretaceous shales or limestones, while the western edge is formed entirely of Cretaceous shaley slates and limestones, and metamorphic pre-Cretaceous rocks do not appear.

Once the depression was formed, deposition began in it. In the northern arm, where the Río Salado now flows, limestones and compact conglomerates were formed in piedmont and alluvial fan deposits, as well as travertines. In the southern part of the Cañada, the Cuicatlán series were deposited. Lowest of the Cuicatlán series are red and gray sandstones, with some gypsum. Above these are massive dark red conglomerates. These are in thick, banded layers. A matrix of dark red ferruginous sandstone binds gneiss, slates, quartz, and schist fragments that range in size from cobbles to boulders 1 to 2 m in diameter. All of the fragments are at least somewhat rounded. They seem to represent deposition in an annually-flooding desert river, analogous to, but on a larger scale than that of the present Río Grande. The Cuicatlán series is deposited unconformably on the pre-Cretaceous metamorphic rocks and apparently represents deposits formed after all the Cretaceous material had eroded off, and the metamorphic material that lay beneath the Cretaceous rock was being attached, contributing the rocks in the conglomerates. The Cuicatlán series is inclined slightly towards the east.

After these tertiary deposits were formed they, in turn, were faulted. The most dramatic evidence of this faulting is the Cantil de Cuicatlán, a 300 m cliff that rises on the eastern side of the Cañada, behind the towns of Cuicatlán and San Pedro Chicozapotes. At the base of this cliff is the junction of the gray sandstones of the lower element of the Cuicatlán series with the red conglomerates that form the cliff. The lower gray sandstones appear in the low hills and mesas that extend out from the cliff, as well as forming the low hills on the other side of the river at Valerio Trujano. Barrera suggests that this fault is the continuation of one noted at Tecomavaca.

Barrera argues that this depression was an undrained basin for a considerable time, until the headwaters of the Papaloapan cut headwards through the Sierra Madre Oriental and by river piracy drained the Cañada into the Gulf of Mexico. I cannot accept this interpretation. First, one would expect extensive salt deposits in an interior basin this large. While the Río Salado, as its name indicates, is somewhat salty, the larger affluent, the Río Grande, is not salty, nor are the Cuicatlán series, which represent the deposition at the time that this valley was supposedly a closed basin. Secondly, the Cuicatlán series, with its rounded cobbles and boulders, seems to represent deposition in a seasonally-flooding torrential river, rather than in a long, narrow lake basin. Finally, the size of the mass of the Sierra Madre Oriental and the depth of the canyon of the Santo Domingo where it exits through the sierra, makes river piracy seem inadequate as an explanation.

It is clear that the course of the rivers Salado and Grande are structurally conditioned by the north-south faulting that formed the Cañada. However, it seems more likely that the river early on cut through the Sierra Madre to form its exit to the gulf. Tectonic uplifting of the whole region may have caused rejuvenation of the river, but the river was able to cut down as the mass lifted up, maintaining its exit to the Gulf, and forming the canyon of the Santo Domingo. However, the more resistant metamorphic and crystalline rocks of the Sierra Madre Oriental have formed a secondary base level for the erosion of the Cañada, as they resist the downcutting of the river more than the tertiary conglomerates, limestones, and sandstones. As a result, the floor of the Cañada is relatively flat to Quiotepec, where the Santo Domingo cuts down steeply to the coastal plain of the state of Veracruz.

My interpretation is to a certain extent speculative. I am not a geologist and have not visited the canyon of the Santo Domingo, nor could I get access to large-scale maps or geological maps of the region of the canyon of the Santo Domingo. However, I feel my interpretation is more in accord with known facts and accepted geological principles than an interpretation based on grand-scale river piracy.

The Cuicatec arm of the Cañada, which is the longer part of the Cañada, is drained by the Río Grande. The Río Grande is formed by three major affluents. The Río Tomellín joins it just south (upstream) of Cuicatlán. The canyon of the Tomellín becomes quite steep and narrow just above where it joins the Río Grande. The present railroad bed follows the canyon of the Tomellín up out of the Cañada to Oaxaca.

South of the Tomellín, at El Chilar, still in the broad part of the Cañada, the Río Grande splits into two branches, one of which, the Grande, has its headwaters to the southeast in the Sierra de Juárez. The other main branch, the Río de las Vueltas, flows nearly due north from just over the watershed from the Etla arm of the Valley of Oaxaca. Up this canyon lay the route of the old Camino Real from Mexico City to Oaxaca (Fig. 1).

The Cañada portion of the ecosystem is the floor of the canyon of the Río Grande and the Río de las Vueltas where flat land allows settlement and where small tributaries are available for irrigation. There are three major wide basins along the Río Grande. The largest and most important basin stretches from Cuicatlán to Dominguillo, about 20 km long and almost 10 km wide at the widest point. The other two basins are at Atlatlauca and La Unión, below the Peña de Ejutla, and at Quiotepec. In between these, the canyon closes in. In the flood season, the river stretches from cliff to cliff in the narrower sections.

Quiotepec draws its irrigation from small Río Chiquito de Quiotepec, and, to the south, the Cacahuatal, which enters the Grande between Quiotepec and Cuicatlán. Cuicatlán draws its water from the Río Chiquito de Cuicatlán, which cuts a deep gorge down through the cliffs of the Sierra Madre Oriental between the two highland towns of Concepción Papalo and Reyes Papalo. San Pedro Chicozapote, only a few kilometers south of Cuicatlán, draws its irrigation water from another Río Chiquito. Across the Río Grande from Cuicatlán, Río Valerio Trujano irrigates with the waters of the Apoala, which descends from the Sierra Mixteca and joins the Grande north of the mouth of the Tomellín. Dominguillo draws its water from the Río de las Vueltas, and from another Río Chiquito between Dominguillo and El Chilar. Apparently the Río de las Vueltas is small enough at Dominguillo that irrigation works survive the annual floods. Atlatlauca and La Unión irrigate

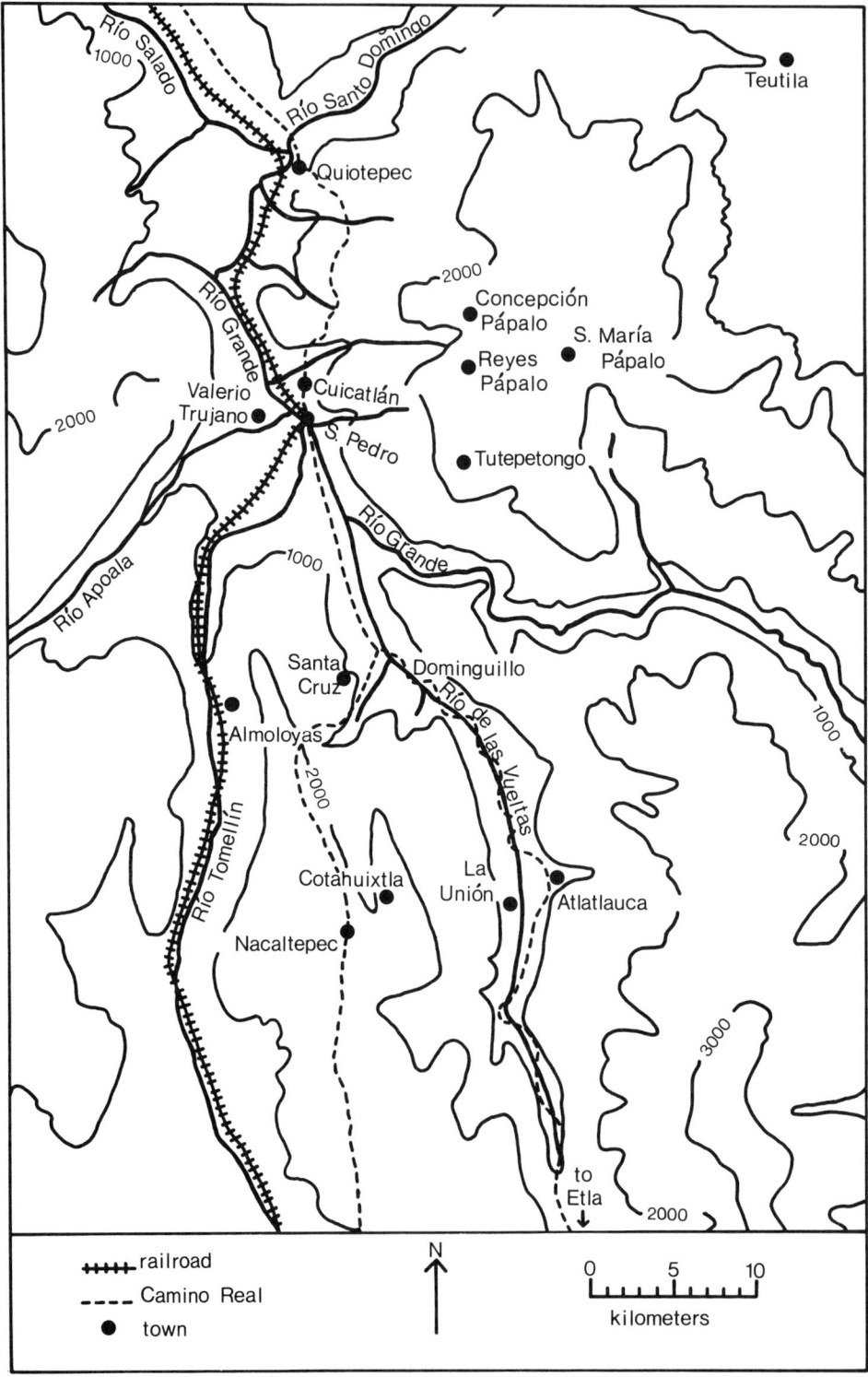

Figure 1. Travel routes through the Cañada.

with small tributaries that descend on each side to the Río de las Vueltas.

The large mountain mass of the Sierra Madre Oriental to the east of Cuicatlán removes much of the moisture from the air before it reaches the Cañada. The Cañada receives on an average less than 300 mm of rain annually. The maximum rainfall usually falls in June, while in January and February no appreciable rainfall occurs. However, in exceptional years, such as 1941 and 1969, as much as 300 mm —the average expected for the entire year— may fall in a single month. In such years disastrous floods cut off communication out of the Cañada, washing out the road and railroad, and carrying off riverside fields. In 1969, when I was in the Cañada, railroad cranes dropped 1-ton boulders in the river to shore up the embankments, only to have them carried away by the river.

The Cañada is well known throughout Mexico for its heat. The average annual temperature is 24.5° C, with a maximum of 43°C and a minimum of 6°C. The greatest heat is in April and May before the rains begin, and the least amount is in the months from October to February. However, the average in the warmest month (May) is 26.5°, while in the coldest (February) the average is 22.6°, not a large range. Since the Cañada is a low basin surrounded by high mountain masses, it forms a cold air sink and gets quite cool at night.

Because of the hot, dry climate, the most common vegetation in the Cañada is low, thorny forest. This is composed of mixed thorny trees and large cacti. Among the trees are palo verde or mantecoso (*Cercidium praecox*), with its bright green bark; mesquite (*Prosopis juliflora*); and quebracho (*Acacia unijuga* Rose). These are low trees 6 to 19 m high. Another low tree found farther up the slopes is the pochote (*Ceiba parvifolia*) which fruits in cottony-looking pods that have a fiber-like kapok and edible seeds. Mixed with these are the largest of the organ cacti, the cardón (*Lemaireocereus weberi*). Other large cacti are the pitahayo (*Lemaireocereus bruinosus*), and pitahayo viejo (*Cephalocereus chrysacanthus*). Smaller cacti include the quiotilla (*Escontria chiotilla*), and nopal (*Opuntia* sp.). In and around Cuicatlán, because of extensive pasturing of goats, there is often not much underbrush under these, and one can walk through park-like aisles between the low trees and cacti, in open shade, only occasionally having to duck under low branches. However, farther from the town, and up the slope from the floor of the Cañada, the undergrowth becomes more closed, with sparse grass (*Pentarrhaphis polymorpha*), the nettle-like mala mujer (*Jatropha urens* L.) and small barrel cacti (*Mammillaria*). On the cliff above Cuicatlán one is struck by the extensive cover of pie de cabra (*Solanum amazonium*), with its pretty purple flowers, until one tries to walk through these waist-high plants and finds out that they are quite spiny.

On the gravel bars along the edge of the Río Grande the vegetation is more sparse because of the annual flooding. Characteristic is the palo de agua (*Asianthus viminalis* H.B.K.) a willow-like tree with soft spongy wood. True willows (*Salix chilensis*) also grow along the river. Small specimens of the thorny scrub also invade the edges of the river. Along the edge of the canal from the Río Grande that led to the sugar cane fields of La Iberia, I saw a large tree which I identified as *Pithecolobium dulce*. The specimen I observed was 10 to 15 m high and 2 to 2.5 m in diameter, but notwithstanding its size, it was later carried off without a trace in the floods of 1969.

In the bottoms of the canyons of the little tributary rivers that supply the irrigation systems, and on the somewhat larger, but still restricted canyons of the headwaters of the Río Grande and the Río de las Vueltas, one finds a more humid microenvironment, with lusher tree and brush vegetation. While the sides of the canyons will still be covered with the same thorny scrub and cactus previously described, the canyon bottoms will support large trees, such as the chupandilla (*Crytocarpa procera* H.K.K.), which has a sour but tasty fruit with a large stone and grows to 15 m long. I describe them as long because they frequently grow out over the canyons, with twisted branches, rather than up. Another tree that looks like chupandilla with its white bark and long twisted roots is the amate, also called higo, or capahuico, which is a *Ficus*, and, I was told, is not used for anything, although in other parts of Mexico its bark is used to make paper. Carrizo, or river cane, often grows right on the river's edge. High on the Río Chiquito de Cuicatlán is a single enormous chicozapote (*Achras zapota* L.), which may be left from a small planting at some previous time. Other trees found are *Pithecolobium dulce* and *Arundo donax*.

The thorn scrub grows up to about 1200 to 1400 m above sea level on the eastern slope of the Cañada. At 1200 m the first oaks begin to appear. The first temporal fields appear at 1600 m. The lower oaks are *Quercus glaucophylla*, *Q. glauccides* and *Q. Liebmannii*. These are low stunted oaks, that were pointed out to me by a Mexican botany student as curious trees that drop and replace their leaves every year. Higher up, *Quercus consperza* grows to 10 to 15 m high, and even higher pines mix with the oaks or form pure stands. Some of the pines are *Pinus montezumae* and *P. oocarpa*. At present the forests above the Cañada are being managed by the Papelería Tuxtepec, with nurseries and controlled cutting. Oaks are being deliberately selected against so the stands of huge pines may be man-influenced. A number of epiphytes, lichens, and mosses are seen on the trunks of the oaks and pines in this more humid environment.

A number of game animals that have long since been driven out or exterminated in the Valley of Oaxaca are still fairly numerous in the Cañada. Deer and jabalí are still hunted, although they are hunted more for sport than for a major source of meat. A number of large predators can still be found. Coyotes are fairly numerous, and occasionally puma, jaguar, and ocelot can be found. Smaller animals include the coatimundi and mapache (raccoon). Vultures, hawks, and eagles live on the cliffs above the Cañada, or in the moist side canyons. Chachalacas and doves occasionally are shot and eaten. Black iguanas, which live in the towns, often in the roofs of the houses, are not eaten, as they are said to eat dung, but the green iguana, which lives in the trees and cliffs along the rivers, is considered a delicacy. Small lizards live all through the thorny scrub. Various snakes, ranging from 1.5 m-long water snakes through rattlesnakes, to coralillo (probably the non-poisonous king snake) live in the Cañada. In the rivers are huile, truchita, mojarra, and pez bobo, as well as some sort of large crab, of which I saw a claw about 20 cm long. Fishing is also primarily a recreational activity rather than a serious economic endeavor.

The basic Mesoamerican crops of corn, beans, and squash are grown in the irrigated fields of the Cañada, as well as in the highland *temporal* fields (unirrigated). However, the Cañada towns are to a large extent also cash-crop producers of fruit and truck produce.

Plants native to America are chicozapote (*Achras zapota*), zapote negro (*Diospyros ebenaster*), ciruela (*Spondias purpurea* L.), jitomate (*Solanum lycopersicum*), papaya (*Carica papaya* L.), jícama, avocado (*Persea americana*), cacao (*Theobroma cacao*), chile (*Capsicum*), and anona (*Annona* sp.). Many of the important crops of the present day Cañada, however, represent post-Hispanic introductions. One of the most important is the mango (*Mangifera indica*) which in various varieties, especially the manila, is one of the prime crops of the area. Another very important introduction is the sugar cane (*Saccharum officinarum*). Just outside Cuicatlán, the hacienda La Iberia used to produce alcohol from the sugar grown there. Now the cane is carried north to Tehuacán in railroad cars to be processed. Coconuts, bananas, rice, and lemons are also grown, but to a lesser degree.

In the highland towns corn and beans are grown as staples. In addition, walnut trees are harvested, and the nuts are brought down to Cuicatlán to be sold. Some coffee is also grown in the highlands, and some is traded through Cuicatlán. Coffee was in vogue at the end of the nineteenth century, but several coffee haciendas listed as being in the Distrito of Cuicatlán before the Mexican Revolution have disappeared now. In addition, logging goes on in the highlands, both on a small scale by the Cuicatec of Reyes Pápalo, who cut and axe out boards and roof beams by hand, and on a much larger scale for paper pulp by the Papelería Tuxtepec.

The population of the meztizo towns on the Cañada floor is the usual mixture of Spanish and Indian people, with some recent Spanish economic immigrants (*gachupines*) forming an economic upper crust. One is occasionally startled to see people with gray or green eyes; this is locally explained by the story that a considerable contingent of French are supposed to have gotten lost in the Cañada on a foray toward Oaxaca during the French intervention. In addition, especially in the towns of Valerio Trujano and San Pedro Chicozapotes one finds considerable numbers of what Esteva Cayetano (1913:97) describes as "familias pertenecientes a la raza etiópica." These people, who are still quite distinctly negroid, are the descendants of slaves brought in to work on sugar haciendas. The town of Valerio Trujano (whose name was originally from the Hacienda de Guendalain), according to the residents, was renamed for the colonel who freed the slaves there.

The present-day Cuicatec live in the highland towns, such as the five Pápalos. Atlatlauca, high on the Río de las Vueltas, is still partly Chinantec speaking. El Cacique, in the mountains south of Cuicatlán, is a settlement of mestizos brought in from another part of the country and set down in an otherwise completely Indian sierra. Recently a number of people have been moving down from the sierra to Cuicatlán, reportedly to escape communal labor obligations in their own towns. These people are extending settlement up the talus slope above Cuicatlán.

The Cuicatec are bordered to the north by the Mazatec in the sierra and Nahuatl speakers in the lower Valley of Tehuacán. To the south, the Chinantec begin at Atlatlauca. To the west, the eastern boundary of the Mixteca Alta is the Cañada. Some of these people, especially the Cuicatec, come down to Cuicatlán to trade. A regular feature of the Saturday market in Cuicatlán are Nahuatl women from Chilac and other towns in the Valley of Tehuacán who sell vegetables and small articles of plastic.

The Cañada has always formed a major natural north-south route of communication. The old Camino Real followed the Cañada up the Río de las Vueltas to Atlatlauca, and across to San Juan del Estado, in the Etla arm of the Valley of Oaxaca. The present road leaves the Valley of Oaxaca at Telixtlahuaca and enters the Cañada at Dominguillo.

A narrow-gauge railway was put through the Cañada, ascending through the canyon of the Tomellín to Oaxaca. In the 1950s this was replaced by standard-gauge track, probably in part due to most of the railway having been carried off in the disastrous floods of 1944.

Finally there is a lumber road up the cliff and the mountains, from Cuicatlán to the forests behind Concepción Pápalo. The road was constructed at considerable expense by the Papelería Tuxtepec and had to be substantially rerouted after the rains of 1969 caused land slippage that removed a half-kilometer of the road below Concepción Pápalo.

Chapter 2

The Cuicatec Ecosystem: An Historical Reconstruction

The Cuicatec ecosystem was defined as the area inhabited by Cuicatec speakers at the time of the Spanish Conquest. Only historical sources can identify which towns spoke Cuicatec at the time of the Conquest. The same sources allow us to reconstruct the Cuicatec ecosystem as it was at the time of the Spanish Conquest.

Historical Sources

The early historical sources fall into several categories. Some are official government records, such as tributary records and *relaciones* made at the request of the government, so that they might administer the colony. Others were from the ecclesiastical authorities (in practice there was considerable overlap between civil and ecclesiastical authority) for use in church administration. Finally, there is the information that comes from diaries and reports of travelers passing through the Cuicatec area. I list below the more important of these sources, with some comments on the kind of information contained in each of them, and, in passing, the way I have used this information.

Codices

Codex Mendocino

This codex is a list of the tribute paid to the Aztecs (the "Culhua-Mexica"). Barlow's (1949) analysis includes material from other early sixteenth-century sources. Two major pieces of information can be derived from it. They are the nature of the tribute the towns paid (and hence what items were available, and roughly what the relative wealth of the towns was at that time) and the political chain through which they paid their tribute, which tells us about the political structure of the larger Mesoamerican entities.

Codex Fernando Leal and Codex Porfirio Díaz

These two codices are copies of the same Cuicatec codex. I have examined facsimiles of these codices but feel incompetent to make any meaningful analysis of them. I join Eva Hunt (1972) in hoping that someone will some day devote the time to these codices that they deserve.

Spanish Sixteenth-Century Sources

Libro de las tasaciones (El Libro de las Tasaciones de Pueblos de la Nueva España, Siglo XVI)

This source is a compilation, town by town, of the tributes of individual towns together with the changes that took place year by year in the *tasación* (tax assessment, or tribute assessment) from 1530 to 1570. It is an important source because it records the tributes levied on individual towns over a period of time. All the towns covered follow a particular pattern.

The first tribute records are usually from the 1530s or 1540s. Tribute was mixed; part was paid in gold or silver, and the rest in produce, skins, feathers, *chalchihuites* (green stones), or the like.

The second set of tributes usually changed from mixed produce to gold. From then on two things happened to the assessment. The assessment in gold dust was changed to *"oro común,"* or to its equivalent in silver *reales*. For example:

> Conmutación del oro—En la ciudad de México, [a] trece días del mes de octubre [de] mil quinientos cincuenta y dos años, estando en acuerdo los Señores Presidente y Oidores . . . a pedimento de los indios de Papalotiquipaque, que están en cabeza de su Majestad, conmutaron el oro en polvo que los dichos indios son obligados a dar conforme a la tasación, a que de aquí adelante lo paguen en reales de plata a razón cada peso de nueve reales, vista la fe de la Contaduría de a como salían los quilates del dicho oro, y que a este respecto de nueve reales de plata cada peso paguen el dicho oro los indios del dicho pueblo y los Oficiales de su Majestad lo reciban, y que esta conmutación se asiente al pie de la tasación y se tome la razón en los libros de la Contaduría. [*El Libro de las Tasaciones* 1952:285]

Tributes in later assessments were often reduced or forgiven.

> (*Al margen:*) Tututepetongo.—En dos de mayo, mil quinientos cuarenta y ocho, atenta cierta información tomada a pedimento de los de Tututepetongo, se mandó que del tributo rezagado den la mitad, y que por tiempo de cuatro años den cada año ciento y veinte pesos de oro común en cada un año, sesenta cada seis meses y cumplidos los cuatro años tornen a cumplir la tasación. [*El Libro de las Tasaciones* 1952:549]

and later:

> (*Al margen:*) Tututepetongo.—En la ciudad de México, catorce días del mes de mayo de mil y quinientos y cincuenta y cinco años, vista la información por los señores Presidente y Oidores de la Audiencia Real de esta Nueva España, tomada a pedimento de los indios de Tututepetongo sobre que no pueden cumplir los tributos en que están tasados, atento lo que por la dicha información consta y la calidad de la tierra y cantidad de gente que hay en el dicho pueblo, dijeron que por tiempo de ocho años primeros siguientes que corran y se cuenten desde hoy dicho dia en adelante, los naturales del dicho pueblo den a su encomendero en cada un año, ciento y nueve hanegas de maíz y ochenta pesos de oro común; cuarenta pesos cada seis meses y no otra cosa alguna, y que este dicho maíz y pesos de oro se reparta entre los naturales del pueblo, repartiendo la mitad menos al viudo y viuda y soltero de por casar, que al casado, y que esta moderación se asiente al pie de la tasación. . . . [*El Libro de las Tasaciones* 1952:550-51]

If the tributes were not reduced, tributes that were not paid were forgiven, effectively lowering the tributes:

> (*Al margen:*) Remisión y moderación por un año.- En veinte y nueve de agosto de mil y quinientos y cuarenta y siete años, a los indios del pueblo de Papalotiquipaque, se les remitío los sesenta pesos de oro que restaban, debiendo de los tributos corridos para que no sean obligados a los dar, y que por tiempo de un año, primero siguiente como eran obligados a dar en cada tributo ciento y treinta pesos de oro en polvo den tan solamente noventa pesos del dicho oro. [*El Libro de las Tasaciones* 1952:285]

Finally, in the 1560s there was a major revision of the tribute. Not only did the amount of tribute jump considerably above the previously listed tribute, but the tribute was (almost always) listed only in corn and gold. The number of tributaries from whom this was to be raised was listed, or implied by a statement about the amount that each tributary must give. In addition, a statement was usually made as to what constituted a tributary, who could not be counted as a tributary, and that widows and widowers counted as half-tributaries:

> En la ciudad de México, veinte y un días del mes de junio de mil y quinientos y sesenta y seis años, los señores Presidente y Oidores de la Audiencia Real de la Nueva España, habiendo visto esta cuenta y visita que fué hecha del pueblo de Tepeucila, que está en la Real Corona, atento lo que por ella consta y parece, y la cantidad de gente que se halló en el dicho pueblo y sus sujetos, siendo presentes los Oficiales de su Majestad, dijeron: que mandaban y mandaron que los dichos indios den de aquí adelante, por todo tributo, en cada un año, trescientos y sesenta pesos de oro común, por los tercios del año, puestos en la cabescera del dicho pueblo, de lo cual haya y lleve su Majestad doscientos y ochenta y ocho pesos de oro, y los setenta y dos pesos restantes, queden y sean para la comunidad del dicho pueblo, lo cual se meta en una caja de tres llaves, que la una tenga el Governador y la otra un Alcalde y la otra un Mayordomo, y presentes todos tres y no de otra manera, se saque lo que se ha de gastar y distribuir en cosas tocantes y convenientes a su República y por de ella, de lo cual haya cuenta y razón para la dar cada que les sea pedida y demandada. Y para pagar el dicho tributo, se reparta a cada tributario casado en todo el año diez pesos de plata, y al viudo o viuda, soltero o soltera que viviere de por si, fuera del poderío de sus padres y tuviere tierras, la mitad, y no se les pida, lleve ni reparta más tributo para ningún efecto, so las penas de las ordenanzas, cédulas, y provisiones de su Majestad, so las cuales dichas penas no se cobre el dicho tributo ni otro alguno de los mozos solteros que estuvieren con sus padres, en el entretanto que no se casaren o salieren del dicho poderío, aunque vieren, y esto guarden por tasación y se asiente en el libro de las tasaciones y se tome la razón en los libros de la Contaduría de su Majestad, y que sea a cargo de los dichos Oficiales, préveer lo necesario al ornato del culto divino del dicho pueblo y sustentación de los religiosos que tienen a cargo la doctrina y conversión de los naturales dél, y así lo pronunciaron y mandaron. [*El Libro de las Tasaciones* 1952:407-08]

Thus the *Libro de las Tasaciones* covers the progress of the transition of the tribute system from the Aztec to the Spanish system. I am not primarily interested in this process of adaptation between the two systems of government but rather in what the tribute assessments themselves can tell me about the populations in the towns assessed. However, to evaluate these tributes one must understand the processes of change.

It would be possible to estimate the population at the time of the Spanish Conquest by projecting the curve of decreasing population in the second half of the sixteenth century backwards and upwards. I did not do this for two reasons. First, I felt that the small sample for the Cuicatec region did not justify this kind of pseudo-mathematical operation. Secondly, I was not entirely comfortable with the assumptions that underlie this kind of extrapolation. These assumptions primarily are those of the shape of the curve on the extrapolated part, and that population was the only factor that affected the tribute charged.

It is enough to point out that the *Libro de las Tasaciones* clearly indicates decreasing population down to the time of the post-Valderrama (after 1560) tributary statement. However, it is nice to have one source in which one can see the process of lowering of assessment for towns. Since the same officials were involved, the same criteria were applied for each entry.

Suma de Visitas

The *Suma de Visitas* is one of the most complete sources of tribute and tributaries for the sixteenth century in Mexico. Conceivably some relationship between tribute per tributary could be approximated by comparing the figures for tributaries in the *Suma de Visitas* (dated to around 1548) with the nearest year's tribute listed in the *Libro de las Tasaciones*. For the reasons stated above, I have preferred not to extend the data that far.

The *Suma de Visitas* gives population in several categories. Sometimes population is in terms of *personas*, in which case an estimate of the total population could be derived by multiplying this figure by 9/10. More often in the data from the Cañada, the data are in the form of *casas*, *hombres casados*, *tributarios*, or *tributarios* or *hombres casados* and *muchachos de menos de quince años*. Occasionally one sees totals of *hombres y mugeres* and *muchachos de quince años o menos*. Where the latter is the case the total is multiplied by 9/10 to get the total population. Where one has *casados* and *tributarios* (equivalent terms, see arguments in Cook and Borah [1960]) one multiplies these by 2 (for the females) and adds the *muchachos de quince años de abajo*. These totals approximate the totals got by multiplying the number of tributarios by 3.3.

Relacion Breve y Verdadera de Algunas Cosas . . . que le Sucedieron al Padre Alonso Ponce en las Provincias de la Nueva España

This is a journal kept by one of the companions of the Padre Ponce on a trip he made down through the Cañada and back in 1568 (Ponce 1967). It gives descriptions of the towns through which they passed and a description of the road in the wet and the dry season, since they passed through in both seasons.

Relación de los Obispados del Siglo XVI

This source dates roughly to 1569 (Cook and Borah 1960; Gerhard 1972). It tells what ethnic identity each town had, how many tributaries they had, what tribute they paid, whether they were part of an *encomienda*, and where the priests were in residence, as well as what other towns these priests served.

Relaciones Geográficas of Various Towns

These famous geographical *relaciones* from around 1580 give quite detailed information, not only on the situation at the time of the *relación*, but also what it was like before the Conquest, as it could be reconstructed from the memory of the oldest people then living. They tell us the language of each town. They also tell us the *sujetos* of each town, from which we can infer the political structure and the size of the political unit.

There is one problem in the interpretation of the *relaciones*. Are we to assume that, because a town was a *sujeto* of another, that it necessarily spoke the same language as the *cabecera*? For example, in the *Relación de Atlatlauca y Malinaltepeque* we are told that Atlatlauca was Cuicatec, while Malinaltepec was Chinantec. The two were included together because they were under the same Spanish *corregimiento* and presumably were also together in pre-Hispanic times. The *relación* gives Atlatlauca six *sujetos* and Malinaltepec one. Can one safely assign these *sujetos* to the linguistic groups of their respective *cabeceras*?

There are *relaciones* for Atlatlauca and Malinaltepec, Cuicatlán, Pápalo, Tepeucila (all in *Papeles de Nueva España*, Francisco del Paso y Troncoso, ed., 1905, Vol. IV), Tutepetongo, and Tonaltepec (the latter two included in the *Relación de Guautla*). An unpublished *relación* for Quiotepec is in the University of Texas Library (Gerhard 1972; Cline 1972).

Later Sources

Geográfica Descripción and Palestra Historial

These two were written by Francisco de Burgoa in the seventeenth century from secondary sources. They were primarily concerned with the history of the proselytization of the area by the church but have some historical information not mentioned in the other sources (see Burgoa 1934a, 1934b).

Diario del Viaje que Hizo . . . a la América Septentrional en el Siglo XVIII el P. Fray Francisco Ajofrín, Capuchino

As its title indicates, this is the journal of a trip by Francisco de Ajofrín, a Capuchin monk, through Mexico, including a tour of the Cuicatec region in 1766 (Ajofrín 1959). It includes very good descriptions of much of the area, especially of the roads and paths, as well as ethnic identifications, ecclesiastical boundaries, and botanical notes.

Travels to Guaxaca

Nicolas Joseph Thiéry de Menonville (1812) was another traveler of the eighteenth century who left a description of his travel through the Cañada, as well as some botanical descriptions.

Relaciones del Siglo XVIII

I have xeroxes of transcribed copies of these from the manuscripts of Paso y Troncoso, loaned to me by Eva Hunt. I believe they are the same as the *relaciones topográficas* described by Gerhard (1972). They are not as complete as the sixteenth century *relaciones*, but they do state which towns were subject to which, in what climatic zone each town was located, and what crops each town grew.

Theatro Americano

This is a summary of the relaciones of the eighteenth century, compiled at that time by José Antonio Villaseñor y Sánchez. It describes the location of towns, population, and commerce, as well as ecclesiastical organization. It tells how many residents are Indians and how many are Spaniards, and it identifies the political jurisdictions.

Historia General and Historia de las Indias Occidentales

These are two histories written by Antonio de Herrera y Tordesillas (1726, 1730) in the eighteenth century, considerably after the events of the Spanish Conquest. However, he drew on sources that have since disappeared.

Recent Works

In addition to these I have drawn very heavily on two modern analyses of the historic material. Gerhard's work (1972) serves both as a condensation of information and as an up-to-date bibliography.

The second is, of course, Eva Hunt's work on the Cuicatec *cacicazgos* (Hunt 1972). Although the data I present in this chapter parallel Hunt's to a great extent, this apparent duplication is justified, I feel, because my analysis was undertaken from a different point of view, within a somewhat different conceptual framework. This in no way implies any important disagreement with Hunt's fine discussion. The publication of her work has aided me enormously in checking my own attempts at historical analysis.

Definition of the Cuicatec Area

I stated earlier that I would define my ecosystem around the area containing populations that spoke Cuicatec at the time of the Spanish Conquest. This, of necessity, must be reconstructed from historical sources. Obviously the older the source in which a Cuicatec town is mentioned, the better the argument that this particular town was Cuicatec at the time of the Conquest. However, some considerably later sources include linguistic identification and suggest which towns have remained Cuicatec up to and including the present. These sources reinforce and supplement the earlier documents, and their recent date allows us to be more comprehensive.

The information from all the sources on the identification of the Cuicatec towns is summarized in Table 1. From this table we can draw boundaries around the Cuicatec area. The floor of the Cañada is the western boundary of this area. The mountains that rise to the west of the Cañada are inhabited by the Mixtec. To the south, the boundary of the Cuicatec territory goes to the top and just over onto the eastern slope of the mountains that form the eastern side of the Cañada. To the north, the boundary seems to have been where the Río Grande joins the Río Salado and forms the Río Santo Domingo, which cuts through the mountains to form the Papaloapan. North of the Santo Domingo are the Mazatec, and on the lower eastern slopes of the Sierra Madre Oriental are the Chinantec.

Some towns in Table 1 are ascribed to more than one linguistic affiliation. Atlatlauca is (mistakenly, according to Eva Hunt [1972]) classified as Chinantec by Herrera. Chiquihuitlán and Quiotepec are said to be half Mixtec or Mazatec. Teutila is said to be Cuicatec but was the head town for an area that was partly Mazatec and Chinantec. Tonaltepec and Tutepetongo, although Cuicatec, were subjects of Mixtec towns.

In some of these cases we can resolve the question of their linguistic identity by taking the majority opinion. Alternatively, we can assume some of the statements represent mistakes influenced by propinquity to the group to which the town is (mistakenly) ascribed. Another possibility for explaining these different statements is that the towns in question have changed linguistic affiliation over time. Eva Hunt (1972) suggests that there has been a general northward movement of the Cuicatec. And finally, the situation is complicated by the fact that some of the towns are now largely or entirely Spanish speaking, making it impossible to determine from the present population past linguistic affiliations.

For the moment I would simply like to point out that all the towns that have ambiguous linguistic affiliation are the border towns of the region. I feel that this ambiguity is present because the towns are on the borders of the Cuicatec ecosystem. I will discuss the importance of this later.

In general, the present-day speakers of Cuicatec are confined to the highland communities, which even today are fairly isolated. The lowland towns are largely Spanish speaking. In addition, the lowland, Cañada towns are often affected by a large infusion of Negros who arrived to work for sugar plantations, especially in Valerio Trujano (Guendalain) and San Pedro Chicozapotes. Atlatlauca is now inhabited by a few Chinantec speakers. In the highlands, Cotahuixtla and Nacaltepec are largely, if not entirely, Spanish speaking. They are now served by daily bus service from the Valley of Oaxaca. The only Indian speakers in Cuicatlán, Dominguillo, and El Chilar, in the Cañada, are recent immigrants from the highlands. The rest of the population of these towns are Spanish speakers.

The Cuicatec Ecosystem

The boundaries of the Cuicatec ecosystem are those of the area defined above. Within this area, each town forms a unit. It is possible that there are areas between the towns that were substantially untouched by people. The Cuicatec towns can be grouped into two types, highland and lowland, on the basis of the physical environment. The lowland towns are all in the Cañada or in the canyon of the Río Grande and its tributaries, especially the Río de las Vueltas. This canyon ranges in altitude from roughly 500 to 800 m above sea level. Because of the mountain mass to the east, the Cañada is quite dry, and because of its low altitude, the temperature is quite high.

The highland towns get a great deal more precipitation. Temperature varies with altitude, and in some localities snow falls occasionally. Eva and Bob Hunt (personal communication) have pointed out that there is a discontinuity today between the highland and the lowland communities. An uninhabited and little-utilized zone lies between the low Cañada towns and the highland Cuicatec towns. This is because the western slope of the sierra, where this zone between the highland and lowland towns is, is in the rain shadow. This creates a zone between the lowest area where enough rain falls for *temporal* fields (in Concepción Pápalo today the lowest fields are around 1200 to 1400 m above sea level) and the floor of the Cañada (at 500 to 800 m above sea level), which can be irrigated by streams descending from the sierra. This zone is (by definition) too dry for *temporal* (rainfall) agriculture, and the streams descending from the sierra are too entrenched to be used for irrigation in this zone. As a result, settlement is only possible above ca. 1200 m, or on the floor of the Cañada, but not in between.

This discontinuity is more apparent at the northern end of the Cañada where the Cañada floor is lower, and the river sources are closer together than in the southern end. In Atlatlauca, a Cañada town, *temporal* farming is possible within a short distance of the town.

The Cañada is a major north-south travel route. This meant that it was fairly easy to get from one Cañada town to another, and from any one of them north or south to the major centers of Mesoamerica. In contrast, the highland towns, by the nature of the towns' location, are not easily accessible, either to one another, or to areas beyond the Cuicatec area. Because of this, the Cañada towns were always more directly exposed to the mainstream of Mesoamerica than were the highland towns.

The mountains in which the highland towns are found, the Sierra Madre Oriental, rise to an average height of over 2000 m, with peaks as high as 3300 m. While there is no weather station, and hence no weather records from the Cuicatec highlands, I observed that it rains enough to provide crops as well as support a dense oak and pine forest. In the higher parts, occasional snow is not uncommon.

This discontinuity between the irrigated Cañada towns and the *temporal* highland towns can be seen on the ascent from Cuicatlán up the logging

TABLE 1
IDENTIFICATION OF CUICATEC TOWNS

Town Name	Relación Siglo XVI	Relación Obispados, Siglo XVI	Padre Ponce 1568	Relación del Siglo XVIII	Theatro Americano 1745	Ajofrín 1766	Cuadros Sinópticos 1883	Belmar 1901	Belmar 1902	de la Cerda Silva 1942	Hunt 1972* Present Identif.?
Acontepeque	sujeto, Atlatlauca				X						
Alpizagua (Dominguillo)		X	X		X		Cuicatec Name		X		
Atlatlauca Santiago Camotlán	cabecera	X			sapoteco X	X					
San Francisco, Chapulalpa		(?)Ulapa					Cuicatec Name	X Chapulalpa	X	Chapulapa	= Ulapa
San Pedro Chicozapotes Chiquihuitlán, San Juan, Santa Cruz						Chinantec Mixtec, Mazatec			X	half Mazatec	= Tzinacantepec
Coapam de Guerrero Comaltianguyzco Coquiapa	"Coapa" (Pápalo) (Pápalo) (Atlatlauca)										
San Francisco Cotahuixtla					Cotahuiztla	Cuicatec Name			X		
San Juan Coyula	(Pápalo)				X				X		
San Juan Bautista Cuicatlán		X	X	Collula (Cuicatlán) X	X	X, some Spanish	Cuicatec Name				
Cuyaltepec, San Pedro							Cuicatec Name		X	"se habla poco"	= Tlacichuya
Cuyamecalco						X, Mixtec, Mazatec Mixtec Name					
Cuytlaquiztlán Cenaguilla, Venta de la Santo Domingo del Río		X									
Dominguillo (see Alpizagua) Guendalain											
Huitzapa	(Atlatlauca)			(Cuicatlán)					X	X	?location = La Laguna?
Huitziltengo	(Atlatlauca)										= S. An. Pápalo
Iscoatula Santiago Ixtlahuaca	(Tepeucila)								X		= Ixtaltepec
Yztactepexi San Juan Bautista Jayacatlán	(Atlatlauca) (Atlatlauca) Xayacatlán						Mixtec Name				= Llano Verde S.F. Nogales
Xoxoctepeque	(Pápalo)									X	

HISTORICAL RECONSTRUCTION

Town	Pápaloticpac	Manaleatepeque	(Pápalo/Cuicatlán)	Nacantepec	Cuicatec Name				Notes
Santa Cruz Juquila	(Tepeucila)								
Maçapa								X	
S. Agustín Montelobos								X	
Nacaltepec									
Pápalo, Concepción	Pápaloticpac	X	(Pápalo)	X			X	X	
Pápalo, San Lorenzo			(Pápalo)	X			X	X	= Yepaltepec
Pápalo, Santos Reyes			(Pápalo)	X	Cuicatec Name		X	X	= Picutla
Pápalo, Santa María			(Pápalo)		Cuicatec Name		X	X	
Pápalo, San Andrés			(Pápalo)		Cuicatec Name		X	X	= Iscoatula
Picutla	(Tepeucila)								= S. María P.
Santiago Quiotepec			Mazatec (Cuicatlán)	Zapotec			X	X	Mazatec
Soluta, S. Bartolo								X	
Thecomastlagua			(Cuicatlán)		Tecomaxtlahuaca				
Tecpanapa	(Pápalo)								
S. Andrés Tetilalpam					Cuicatec Name	Tetilalpam	X	X	
Tepeucila	X	X	(Pápalo)	X	Cuicatec Name		X	X	
Teponaxtla, S. Juan	(Tepeucila)		(Pápalo)	X	Cuicatec Name		X		(Tepeucila)
Tequecistepeque		X							
Teutila		X		Mazatec	Cuicatec Name		X	X	
Tlachichuya	(Tepeucila)								
S. Sebastián Tlacolula	(Tepeucila)		(Pápalo)		Cuicatec Name		X	X	(Tepeucila)
Santa María Tlalistac					Cuicatec Name		X	X	
S. Juan Tlalixtlahuaca							X		
Tlaxila, S. Catarina							X		
Tonaltepec, S. Juan					Cuicatec Name		X		
Tutepetongo, S. Francisco	Tanatepeque (Guautla)	X							= Tanatepeque
Tututepetongo (Guautla)	Tututepetongo (Guautla)	X	(Pápalo)		Cuicatec Name		X		
Xayacatlán (see Jayacatlán)									
San Lorenzo Yepaltepec	(Pápalo)							X	= San Lorenzo Pápalo
S. María Yololtepeque									
Yolutla	(Tepeucila)								
Zapotitlán					Chinantec Name				(C. Pápalo)

X = Statement that this town is Cuicatec.
( ) = *sujeto*, *barrio*, or *estancia* of town in parentheses.
*While Eva Hunt discusses almost all these towns, I have included items in this column only when it says something beyond what is already contained in the table.

road to Concepción Pápalo. As soon as one gets above the irrigated land, one is in xerophytic scrub. This vegetation continues up to about 1200 m above sea level, where the first sparse oaks begin to appear. At 1600 m one sees the first fields. Oaks still predominate, although some pines begin to appear. Finally, higher up are pure stands of large pines. The towns occur at intermediate altitudes. Concepción Pápalo is at 2280 m, and Santa María Pápalo is at 2040 m by altimeter. Tepeucila, according to the topographic map, is at about 2200 m.

These highland towns were to a large extent agricultural, growing the usual Mesoamerican crops of corn, beans, chiles, and squash. Some of the fields of the highland towns would have been planted only once a year because they were high enough to suffer from frost in winter. Most of the highland agriculture was, and is, rainfall or *temporal*. Concepción Pápalo today uses small springs to irrigate a few fields, but most of the fields depend on rainfall. The *Suma de Visitas* states that Papalotiquipaque (Concepción Pápalo) had some irrigation, but the *relación* for the same town says,

> En las quebradas que ay entre estas sierras pasan algunos arroyos que van a baziar a vn rrio grande que pasa çerca de si que se dize el Rio de Albarado: naçen muy çerca de la cabeçera de este pueblo y porquestan ahoçinados no se pueden sacar ni tiene dellos provecho. . . . [*Relación de Papaloticpac* 1905:92]

Some products that would be limited to the highlands may have been important in their economies, in addition to subsistence staples. First, the highlands grew maguey to produce pulque. This kind of maguey will not grow in the Cañada, so the trade in pulque may have been important. Both Santa María and Concepción Pápalo grow pulque today. The *Relación de Guautla* (1962:12) tells us that "en tiempo de su gentilidad" in Tutepetongo the Cuicatec danced in front of the idols drunk with a wine that they made from maguey or from ciruelas:

> En tiempo de su gentilidad . . . sacaban sangre de la lengua y orejas; e la ofrecían al ídolo, é acabado bailaban delante dél con guirnaldas de flores en las cabezas, imbriagados hombres y mujeres con un género de vino que entre ellos hacían de maguey o de cirgüella de la tierra.

Guautla, which was Mixtec, not Cuicatec, but had Cuicatec *sujetos* after the Conquest also made "vino" and gave maguey as part of its tribute.

> Este pueblo de Guautla fué en tiempo de su infidelidad de un Señor que se . . . nombraba Ocelotecotle, . . . al cual le tributaban ropa de algodón tejida y en mantas, venados maiz que le sembraban é magueyes, que es un árbol de mucho provecho y granjería, de que sacan miel é otros muchos aprovechamientos. . . . [*Relación de Guautla* 1962:5]

The highland towns were in the oak-pine zone. They built their houses of this wood. There is no mention that they traded this wood with the lowland towns, either as building material or as charcoal. Today the people of Reyes Pápalo drag roof beams (*vigas*) and carry broad boards (*tablas*) and nets full of charcoal on their backs to Cuicatlán to sell. On the other side of the Río Chiquito, the Papelería Tuxtepec spent several millions of pesos to construct an all-weather road to Concepción Pápalo, and to the mountains behind it, in order to bring wood down to the railroad in Cuicatlán for making paper pulp. Since the trees in the xerophytic flora of the Cañada are not suitable for planks or roof beams, this pattern may persist from pre-Hispanic times.

Several sources attest that such trade existed in other parts of Mesoamerica in early times. Cortés, in his description of the market in the second *Carta Relación*, mentions "carbón" (charcoal), "braseros" (vessels for burning charcoal, either as a stove, or for incense) and "leña" (firewood). In the *Suma de Visitas*, for no. 301, Ystepexi, we are told that "la grangeria que ay es en tablazon y vigas." This town was in *tierra fría*, near the Cuicatec area, although it was not Cuicatec. This supports the idea that this trade may have existed in pre-Hispanic times among the Cuicatec towns.

Petate (mat) making, from a kind of palm that grows in the mountains, is another pre-Hispanic specialty that continued on to post-Hispanic times. It was practiced in some highland towns in the Cañada vicinity. We are told that Tanatepeque (Tonaltepec) "tiene por trato . . . vender petates, que son como esteras de España . . ." (*Relación de Guautla* 1962:16).

Copal (the resin used for incense) also came from the highlands. The *Relación de Cuicatlán* tells us that "en los cerros de la comarca ay unos arboles apparrados pequenos, que se dizen 'copales' que echan una goma olorosa que para sahumar perece encienso" (1905:187). Copal was important in pre-Hispanic religion as incense. It did not grow in Cuicatlán.

Tepeucila produced its own salt.

> Ay en las vertientes de este pueblo hazia el rio questa dicho vnas peñas donde mana çierta agua y hazen represa della y se cuaxa y haze sal, lo qual solamente se cuaxa en tiempo de seca

avnque avenidas que an pasado an derrunbado parte de estas peñas, lo qual a lo que dizen a sido cavsa de que sea cada dia menos, y esta salina es del Cacique, porque devo ser posesion antigua. . . . [*Relación de Tepeucila* 1905:98]

This salt spring is also mentioned in the *Relación de Cuicatlán*, which repeats that it was small. "Proveese este pueblo Cuicatlán de sal de los comarcanos, porque en este no ay salinas que la den, y diez leguas de aqui se cria y coje en cantidad y en el pueblo de *Tepeucila* se coje alguna, avnque poca" (1905:188). However, at least one other Cuicatec town, Tanatepeque (Tonaltepec), depended on Tepeucila for its salt. "No tienen salinas para se aprovechar, y van á comprar la sal para su sustento al pueblo de Tepeucila, que está en el Real Corona, cinco leguas deste, camino áspero . . ." (*Relación de Guautla* 1962:16). This may mean that the salt source in Tepeucila was a little larger than they claimed.

Hunters used the zone between the highland and lowland zone. This contributed to the food supply, as well as possibly supplying material for trading. For Pápalo, the *relación* says: "Su ordinaria comida era . . . algunos rratones que caçavan en el campo, . . . avnque los Señores y gente principal comian gallinas de la tierra y benados y codornizes, . . ." (*Relación de Papaloticpac* 1905:91). For Tepeucila, similarly,

Su comida ordinaria era tortillas y chile y frisoles y si alguno caçava algun venado o conejo o rraton lo comia avnque por la mayor parte todos lo presentavan a su señor natural y dello les dava alguna cosa, o se lo gratificava en otra cosa de comida, o vestido, porque a los Señores solo era . . . permitido comer gallinas y codornizes y venados y otras caças. [*Relación de Tepeucila* 1905:96]

This theme that meat, particularly of deer, was limited to the *cacique* and the *gente principal* is repeated in several relaciones, although many seem to feel that "rratones" were permitted to the ordinary people.

The *Relación de Tepeucila* tells us that in the time of the Aztecs they paid tribute in "cueros de tigres que conpravan de los pueblos comarcanos y algunas plumas" (ibid.:95). The *Relación de Guautla* says of Xocoticpaque, a nearby Mixtec town, that "los tratos y granjerías que tienen los naturales son . . . mucha caza de venados, liebres é conejos é otra cazas" (*Relación de Guautla* 1962:9). Thus skins and other products of hunting may have been a highland source that could be traded out. Tutepetongo also caught some fish.

The "gallinas de la tierra" were domestic turkeys. These were one of the major pre-Hispanic domesticates and sources for meat. It is difficult to tell whether the other birds were domestic or wild. The language of the *relaciones* is ambiguous. For food we are told they ate "codornizes" (quail). These presumably were wild. Of Tanatepeque (Tonaltepec), we are told that, "Crianse aves de la tierra é de Castilla, palomas caseras é de monte, conejos, codornixes, leones y lobos, gallinas de la tierra y monteses" (*Relación de Guautla* 1962:16). This seems to indicate that turkeys were domestic (which we knew) and also wild, as they are included in the list of wild creatures.

This question of whether birds are domesticated or not is complicated by the fact that birds served for more than food. They were also used for trade, for many things. We have already seen that Tepeucila paid tribute in *plumas*. Tepeucila's name means "hill of the hummingbird place," from *tepetl* = hill or mountain + *huitzi* = hummingbird + *tla* = location (Eva Hunt, personal communication) (and not, as Bradomín [1955] says, hill where hummingbirds are abundant). The *Relación de Tepeucila* (1905:94) says the name refers to

vna sierra questa sobre el propio pueblo donde dizen se crian vnos paxaritos muy pequeños que tienen nonbre «huyçiçiles», los quales crian vna pluma conque se hazen pinturas, la qual en ellas y en los propios paxaros haze vnas aguas o colores diferentes de agradable vista. . . .

It is possible, but unlikely, that these birds were domesticated. Probably the birds were hunted in the wild. Alternatively, it is possible that the birds were captured wild and plucked, then released or imprisoned in cages without being bred in the towns. At any rate, these feathers were an important highland product, at least in pre-Hispanic times.

Two other highland Cuicatec economic activities were gold mining and the raising of *grana* (cochineal). Pápalo had a gold mine which was buried in an earthquake after the Conquest, but before the *relación* was written. Grana, a small red insect which is used to make a red dye, and which grows on nopales (*Opuntia* sp.) was quite important in post-Hispanic times up to quite recently when it was replaced by synthetic dyes. Tutepetongo has "tunales para criar grana" (*Relación de Guautla* 1962:13). In Tepeucila, "En algunas estancias de este dicho pueblo se da alguna grana, porque como es serrania no tienen espaçio ni lugares descanpados para podello senbrar, y ansi es poco lo que cogen" (*Relación de Tepeucila* 1905:97).

The description of the highland towns makes them sound fairly self-sufficient. However, it is clear from the records that they were specialized within a regional exchange system. Two important economic activities of inhabitants of highland towns mentioned in the *relaciones* were weaving cotton and going to the lowlands as migrant labor. This is either mentioned as their *granjería* (trade, occupation) or as what they did to get money to pay their tribute.

We are told that in Pápalo,

> Tienen por contrataçion y granjeria los naturales de este pueblo conprar algodon del dicho Rio [de Alvarado] que se cria en abundançia, y esto hilan y texen sus mugeres y hazen dello naguas y mantas ques el vestido de los naturales, lo qual les viene a conprar españoles y naturales y lo llevan a vender al propio Rio donde se truxo el algodon y a otras partes, y del valor que sacan pagan sus tributos en dineros. [*Relación de Papaloticpac* 1905:93]

The *Relación de Tepeucila* says that,

> No tienen los naturales de este pueblo tratos ni granjerias ni se dan a ser mercaderes y para pagar sus tributos . . . se van a alquilar a los pueblos de la comarca que ay en ellos simenteras de riego. [1905:98]

The Mixtec highland towns of Guautla, Xocotiqpaque, and Xaltepetongo also seemed to fit this pattern. Guautla's *relación* tells us "tienen por trato é granjería . . . hacer ropa de algodón que venden." They got the cotton from the "Costa del Norte"; Xocotiquaque had among other things as its *trato*, "Ropa de la Tierra" (*Relación de Guautla* 1962:9); and Xaltepetongo sold corn, part of which they grew in the following way:

> Diez, digo, dos lenguas deste pueblo, en tierra caliente, está en río que se dice de Cuicatlán, de que los naturales deste pueblo se aprovechan en ir a los vecinos dél á les arrendar tierras, por no las tener ellos, é otro que pasa muy bajo, de que se aprovechan de algunos regadíos. [*Relación de Guautla* 1962:11]

The Cañada Towns

The lowland towns are in the rain shadow of the Sierra Madre Oriental. The annual average rainfall in Cuicatlán is 300 mm, compared to the 600 mm per year that is considered to be the lower limit of rainfall that will allow unirrigated agriculture in Mexico. In Cuicatlán most of the rain falls between June and September. Practically no rain falls between October and May. The temperature ranges from a maximum of 43.0° C (109.4° F) to a minimum of 6° C (42.9° F). The hottest part of the year is from April to June, before the rains. The coolest time is after the rains, from October to February. However, the annual range of average temperatures is really quite small, 3 or 4 ° C, which is considerably less than the diurnal difference in temperature.

The Cañada towns lie at altitudes from about 600 m above sea level at the lower end of the Cañada where Cuicatlán and Quiotepec are located, up to about 800 m at the upper end at Atlatlauca.

Rainfall agriculture is impossible in the Cañada. The lowland Cañada towns were all irrigated in pre-Hispanic times. Cuicatlán is described in the *Suma de Visitas* as "Tierra muy calida y (tiene) un rio; y frutas y cosas de comida se dan muy bien. Es tierra de regadios" (1905:101). Padre Ponce, after coming south from Quiotepec and crossing two arroyos (the Río Chiquito de Quiotepec and the Cacahuatal) describes Cuicatlán in the following manner.

> Es aquel pueblo muy fesco y fértil de fruta, especial de plátanos y de chicozapotes, de los cuales hay muchos plantados en el mesmo camino, orilla de un arroyo que entra en el pueblo y pasa adelante, con que riegan los indios sus milpas y huertas. [Ponce 1967:19]

The *Relación de Cuicatlán* gives the fullest description of Cuicatlán's agriculture. The *relación* states that the "Rio de Alvarado" (the present Río Grande) was not used for irrigation because of its flooding and because "Viene por quebradas y entre sierras asperas, que es causa de que tengan poco provecho para mas quel pescado" (1905:187). But,

> Baxa de vna sierra comarcana a este pueblo vn arroyo de mucha y buen agua, con la qual se rriega todo la arboleda, y avn las simenteras y legunbres que sienbran los naturales de este pueblo, y ansi mismo por la parte del medio dia le desciende otro arroyo de buen agua y mucha, con que se rriega otra parte de frutales y simenteras que los naturales tienen en laderas y mesas que a fecho el Rio Grande. [*Relación de Cuicatlán* 1905:187]

These would be the Río Chiquito de Cuicatlán and the Río Chiquito de San Pedro Chicozapotes.

Irrigation in Dominguillo, or Alpitzauac (Alpizagua), is mentioned by the Padre Ponce in the following manner: "Pasa por medio del agua acequia con que rriegan los indios maizales" (Ponce 1967:20). The present town takes the water off a branch of the Río de las Vueltas, with a masonry canal up against the side of the outcrop of rock above the town. Hence the "por medio del agua."

Atlatlauca was in a valley a quarter of a league by a half a league wide,

> . . . y azia vn lado del dicho pueblo pasa el rrio; de suerte que ba por medio destas dos serranias: ba a dar a rrio de Cuicatlan

y de alli al de Aluarado, quel vno y el otro ba a dar a la mar del norte: lleua siempre cantidad de agua, aunque en partes del ba hondo en tienpo de aguas . . . corre de sur a norte y rriegase con este rrio las sementeras de los naturales, porque como es tierra caliente, los maizales y sementeras de los naturales son de rriego, y ansi se aprovechan todos los yndios del dicho rrio para rriego de sus sementeras, los quales tienen todos por el rrio abaxo. . . . [*Relación de Atlatlauca y Malinaltepec* 1905:172].

Further on, the same *relación* states that,

Ninguno otro rrio ay en este dicho pueblo de Atlatlaucca, ni de Malinaltepeque, mas del questa dicho, el qual corre de sur a norte: ba a dar al de Cuicatlan, questa ocho leguas de aqui; rriegan con el las sementeras para su sustento. [ibid.:173]

This is a contrast with the present situation; the modern town of Atlatlauca is on a small tributary of the Río de las Vueltas that bifurcates above the town and is used for irrigation.

Quiutepeque (Quiotepec) is described in the *Suma de Visitas* (1905:189): "es tierra callente y enferma, tiene vna buena vega de Riego donde se coge maiz y axi y otros bastimentos, es arenosa y salitral . . . tiene la vega dos leguas de largo y media de ancho." This presumably refers to the areas they irrigated with the Río Chiquito de Quiotepec and the Cacahuatal.

With this irrigation the people of Cuicatlán grew, in addition to the omnipresent maize, a lot of fruits and vegetables. Corn was grown with irrigation, "porque de temporal jamas lo siembran como en otros pueblos." Beans, chiles, sweet potatoes, "y otras semillas con que se sustentan los naturales" were grown (*Relación de Cuicatlán* 1905:188). They grew a number of different fruits, including chicozapotes, avocados, red zapotes, ciruelas, bananas (*plátanos*), and even some cotton. These fruits were said to be the best in New Spain: "como es tierra caliente ay en el muchas frutas de la tierra y muy buenas, que se tiene por çierto son las mejores de la *Nueva España*. . . ." (ibid.:187).

In addition to the same fruits Cuicatlán grew, Atlatlauca grew anonas and a fruit called *cuaxinecuiles*, which are pods with a white sweet flesh. Like Cuicatlán, their commerce is based on the surplus and "cash crops" they produced.

Some fishing was done in the rivers. The *Relación de Cuicatlán* says that the Río Grande was not used for anything but fishing. The *Suma de Visitas* says that Atlatlauca had three rivers in which there were a lot of fish, and the *relación* of the same town says they obtained "truchas," although they were small, from the river they used for irrigation (the Río de las Vueltas). Padre Ponce tells us that after his fatiguing journey to Dominguillo, in the heat and among the mosquitos, "fueron los indios a pescar a un río que está allí cerca y hiciéronle caridad de la pesca, que toda fue poca" (Ponce 1967:20).

Some hunting was also done in the lowland towns. The *Relación de Cuicatlán* (1905:186) tells us that "el Señor comia gallinas, benados, conejos, codornizes y carne de ombres y de niños o mujeres, quando la matavan en las gerras" (if this last can be thought of as hunting). We are also told that, "En esta commarca ay cantidad de benados y algunos leones y rraposos, y vnos gatos monteses pintados a manera de tigres. Ay conejos en cantidad, halcones, gavilanes, y otras aves de rrapiña de la tierra" (*Relación de Cuicatlán* 1905:188).

The *Relación de Atlatlauca y Malinaltepeque* tells us more about this hunting. They ate

. . . liebres, conexos y benados. Aunque antihuamente no la comian todos, porque se les bedaua la caça por los Señores . . . comen lagartixas y rratones y otras suçiedades. Antiguamente los macehuales no pudian comer gallinas sino solo los prençipales. [1905:171]

"Lagartixas" may have included iguanas, which are still hunted in Cañada.

The lowland towns also seemed to raise or hunt some birds. The *Relación de Atlatlauca* tells us that their clothing consisted of mantles,

. . . y estas mantas eran listadas de colores, y texidas muchas labores por abaxo. Tenian vna como çanefa hecha de labores y entretexida por ellas plumas blancas y otras colores, y para este efeto criauan vnos paxaros que son de la manera que anadones, saluo que son mas grandes y tiene el pico colorado, que los llaman en cuicateco *dzacha*, y en mexicano *canauctli*. Estas mantas traian los prençipales. . . . [1905:170]

"Plumeria verde y de todos colores" was also part of their pre-Hispanic tribute (*Relación de Atlatlauca* 1905:165).

Cuicatlán grew a little cotton: "Cojese en este pueblo algun algodon de que los naturales haxen rropa para su vestido. . . ." They also got some gold by panning for it in the Río Grande, although this never was very important. "En tiempo de su gentilidad para pagar el tributo a Montecçuma lavavan oro junto al rrio, avnque se allava poco, y en otros arroyos ansi mismo lo buscavan avnque siempre fue trabajoso de hallar" (*Relación de Cuicatlán* 1905:188).

One gets the definite impression that the Cañada towns were less self-sufficient than the highland towns. In particular, the lowland towns seemed to be specialized in a kind of agricultural "cash cropping." Fruit formed an important part of the "cash cropping." The *Relación de Cuicatlán* (1905:188) says that "Tienen por granjeria bender las frutas que crian en este pueblo, porque de toda la comarca se lo vienen a conprar y se lo pagan para llevallo a otros pueblos y en especial a la çibdad de Antequera." This despite the fact that we are told in the *Suma de Visitas* that Antequera (Oaxaca) was two days' journey from Cuicatlán. However, the fact that the reference uses the name Antequera, the Spanish settled city in the Valley of Oaxaca, may indicate that this trade was post-Hispanic, perhaps because of the introduction of the horse.

The *Relación de Atlatlauca* tells us that,

> La contrataçion de los naturales destos pueblos es el maiz que coxen, y de las otras semillas y de fruta que coxen en gran cantidad de la tierra; las quales van a vender y a trocar a los tianguez y otros de otras partes las traen, y este es su trato y granxeria.... [*Relación de Atlatlauca y Malinaltepeque* 1905:175]

The security of the irrigation-based agricultural production of the Cañada towns meant that towns that depended on rainfall often turned to the Cañada for staples when their crops were short. For example, Pápalo went to Cuicatlán in emergencies: "y con esterilidad de tenporales y falta de bastimentos se proveen del pueblo de Cuyctatlan por los rregadios que tiene de mayz y otras semillas y bastimentos" (*Relación de Papaloticpac* 1905:93).

In addition to this trade in agricultural products, which after all, while characteristic of the lowland towns, is not unique to them (Tutepetongo, a highland Cuicatec town, also sold "maiz, tomate, aji" [*Relación de Guautla*] as their trade) the lowland Cuicatecs to some extent may have been merchants through the area. The *Relación de Tepeucila* (1905:93) tells us the "Proveense este pueblo de sal de otros comarcanos que cayn en el camino Real que va a la çibdad de Mexico." This might have included Cuicatlán, although more likely it was Teotitlán del Camino, farther up the Camino in the Valley of Tehuacán. In addition, the *Relación de Atlatlauca y Malinaltepeque* (1905:175) tells us the "proveense.... de Cuzacatlan y Cuicatlan, que son ocho y catorze leguas de aqui." This implies that there were *yndios* from Cuicatlán who went to Atlatlauca to trade salt. Traveling traders and porters are also suggested in the passage about cotton in the *Relación de Cuicatlán* (1905:188): "Cojese en este pueblo algun algodón ... avnque la mayor parte de lo que gastan es del *Rio de Albarado* y de otros a el comarcanos que lo trayn en yndios, questara como beynte leguas de este pueblo."

The role of Cuicatlán as a center for trade and other more or less pan-Cuicatec activity can in part be traced to the fact that Cuicatlán lies on the Camino Real, a natural north-south route through the highlands of central Mexico. The canyons of the Salado and the Río Grande make a natural route through the central mountains, one of three in Mesoamerica—the other two are (1) down the Gulf coast, or (2) through the Morelos valley and then through the Mixteca Alta. Cuicatlán lies in the middle of the route through the lowland Cañada, along the Río Grande. Beyond Cuicatlán, the route split, so that Atlatlauca was only on the dry-season route.

In contrast, all of the highland towns were back in the mountains, and any route that reached them had to work its way through the steep mountains, dense vegetation, and uncomfortable climates, ranging from tropical rain forest through cold pine forest with occasional snow to xerophytic desert scrub, with a shortage of water. These roads had to traverse extreme slopes and dangerous cliffs. These routes connected individual towns, or linked the towns with the Camino Real.

All north-south transit from the Valley of Mexico to Guatemala would tend to pass down the Camino Real, through Cuicatlán. Thus contact beyond the immediate Cuicatec area tended to be oriented first down from the highland towns to Cuicatlán, focusing on Cuicatlán, and then north-south along the lines of the Cañada and the Camino Real, to Oaxaca and points south, and Tehuacán and points north.

The Camino Real had a major dividing point just north of Cuicatlán, in Tehuacán. Burgoa, in the seventeenth century, tells us that,

> Está el sitio de la Villa [de Tehuacán], tres leguas de la Ciudad. Saliendo de ésta parte para el poniente, yendo al Noroeste, es camino real para Mexico, y la Vera Cruz, que se divide en Tehuacán. [1934a, II:8]

We know the route of the Camino Real was used before the Conquest as a major route from several sources. Herrera (1730, Decada III, Libro III, Cap. XV:101) tells us that the batallions of Montezuma passed through Tecomavaca "que esta en el Camino Real de Guaxaca a Mexico" when they were on their way to conquer the people of Zapotitlán. Burgoa (1934, II:11) refers to the Mexicans, finding their way cut off through the Mixteca, attacking the Zapotec from the "Camino Real de los Cuicatecos." Torquemada, perhaps describing the same period of trouble with the rebellious Mixtec, led by Coixtlahuaca, describes how, having been unsuccessful in their attempt to attack the Mixteca directly, they had to make a wide swing around and attacked by way of Guautla (in the Mixteca, not Huautla Jiménez), thus entering the Cañada and crossing the mountains to the west of the Cañada to the Mixteca.

> No hallaron paso, por que ya todos los mixtecas, estaban mui a lo descubierto, puestos en arma, y fueles forçoso, hacer un rodeo mui grande, y de muchas leguas, y llegaron a Huahtlan. . . . [Torquemada 1723, I:208]

We know the route through the Cañada shortly after the Conquest, in 1568, because of the itinerary of Padre Ponce (Fig. 1). Apparently the dry season route was different from the wet season route, because of the difficulty of crossing and recrossing the Río Grande in its flood stage. When Padre Ponce went south to Guatemala, it was the dry season. He entered Oaxaca by passing down the Valley of Tehuacán to Tecolutlán, or Los Kues (the present day Los Cues). From here he proceeded by good road to Tecomahuac (Tecomavaca). Here he crossed two rivers (one of which must be the Salado, and the other the Río Grande, or the combined Río Grande and the Salado, or Santo Domingo). From Quiotepec, the road went up the mountains on the eastern (Cuicatlán) side of the Río Grande and crossed two arroyos (the Río Chiquito de Quiotepec and the Cacahuatal) to Cuycatlan (Cuicatlán). The road departed from the canyon of the Río Grande between Quiotepec and Cuicatlán, because the canyon of the Río Grande was quite narrow there and was flooded by the river in the rainy season. A league past Cuicatlán, the road recrossed the Río Grande (probably above the mouth of the Tomellín, but below El Chilar, but possibly below the Tomellín, since they crossed one river, "dos o tres vezes") and then crossed a few arroyos (small tributaries between El Chilar and Dominguillo) to arrive at Altpizauac, or Dominguillo. At Dominguillo the road climbed the mountains, passing a big dropoff called the Salto del Puerco, across a couple of small arroyos, to the inn at Cienaguilla. At this point they were still in the Cuicatec area. They then passed the Casa de la Seda (San Sebastián Sedas?) and arrived in Quauhxolotitlán (the present day Huitzo). From here they entered easily down the Etla arm of the Valley of Oaxaca, to the city of Oaxaca. In the rainy season the return route of Padre Ponce was the same, except that at Dominguillo the road went up into the mountains on another "Camino que llaman de las Vueltas, porque son infinitas las que en él se dan, para poder salvar el dicho río e innumerables barrancas muy hondas y peligrosas" (Ponce 1967: 58). This route crossed four arroyos and one river, the Tomellín, to arrive directly in Tecomavaca. Apparently the rainy season Río Salado was not nearly so formidable an obstacle as the Río Grande, which in time of floods could not be forded and had to be crossed on rafts.

Other later travelers give us slightly different routes, but they agree roughly on the route of the Camino Real. In 1766, Francisco de Ajofrín visited the Cuicatec area extensively and left a good description of his travels. His route was down the Salado through Venta Salada and San Juan de los Cues, then across the Salado twice and across the Río de Quiotepec (either the Grande or the combined Grande and Salado), to Quiotepec. At this point Padre Ajofrín made an excursion to one side to Teutila, but then he returned to Quiotepec, and the Camino Real "para Oaxaca, Guatemala y otras provincias."

The stretch between Quiotepec and Cuicatlán, which also impressed Padre Ponce, ascended the mountains on the east side of the Río Grande and descended at Cuicatlán. Padre Ajofrín said of this stretch,

> Desde Quiotepeque hasta Cuicatlán es algo áspero, pedregoso en parte, de mucha arena y no menos polvo; no hay agua en todo el camino sino a la salida de Quiotepeque, en un arroyo que se pasa tres veces sin puente ni canoa. Esta tarde padecí la mayor sed y fatiga que he tenido jamás y aun todo el dia fué de suma fatiga, ya por la jornada tan dilatada como por lo fragoso del camino, en que apenas di paso con seguridad ni fijeza. [1959:82]

Further along in the same passage he noted that there are various routes between Quiotepec and Cuicatlán, and suggested that the one which follows the river was the best, although he probably did not realize that this would have been impassable in the rainy season.

From Cuicatlán one went a quarter of a league, then crossed the Río Grande. Ajofrín put the town of San Pedro Chicozapotes on this (west) side of the river, which is incorrect. He then continued through El Chilar, and then, he said, crossed the Río de las Vueltas twice. He probably confused the two small tributaries that enter the Río de las Vueltas between El Chilar and Dominguillo with the Río de las Vueltas. Ajofrín had this to say of the stretch from Cuicatlán to Dominguillo: "Aunque es llano el camino, tiene mucha piedra y arena que le hacen muy pesado" (1959:83). At this point he followed the Río de las Vueltas up its canyon, but he noted the alternative route that Padre Ponce followed, from Dominguillo, to Cotahuistla (Cotahuixtla), to Huiso (Huitzo), and then into the Valley of Oaxaca.

However, Padre Ajofrín instead went up the canyon of the Río de las Vueltas, crossing it 52 times, and noting the number of times one crosses varies from year to year. Following the Río de la Vueltas he passed through Atlatlauca to Jayacatlán. At Jayacatlán he ascended the slopes of the mountains to San Juan del Rey (San Juan del Estado?), where he entered the Valley of Oaxaca. Later Padre Ajofrín returned to Mexico by way of the Mixteca Alta, and at the end of the description of this trip he expresses his preference for the Cañada route as infinitely better, the only advantage of the Mixteca route being that it avoided the Río Grande in flood.

> Ahora, en mi sano juicio, hallándome en Tepeaca, y reflexionando las jornadas tan dilatadas y muchos trabajos que he padecido en la Mixteca, hago proposito firme de nunca jamás tomar este camino para Oaxaca por largo, fragoso, inaccesible y desatinado, y suplico escarmienten en mí los que lean esto; váyanse por el camino real de Theguacan, Cuicatlán, etc, como a la p. 82; es mas breve y no tan aspero. De suerte que tomando este camino hay desde aqui a Oaxaca 75 leguas, y por la Mixteca, 86. Verdad es que se toma este camino de la Mixteca en tiempo de aguas por huir del río Salado y otros caudalosos, pero no faltan ríos por la Mixteca, como hemos visto, y acaso más peligrosos. [Ajofrín 1959:137]

M. Joseph Thiéry de Menonville (1812) was another eighteenth-century traveler. Thiéry de Menonville was a Frenchman who sneaked into Mexico to steal specimens of cochineal and the plants they grew on, in order to smuggle them out and grow them in the French colonies. He followed much the same route as his predecessors, down from Tehuacán through Los Cues to "Aquiotepec," then to "Quicattan" (Cuicatlán) then to Dominguillo, after crossing the Río Grande a league and a half from Cuicatlán. At Dominguillo, he, like Ajofrín, went up the Río de las Vueltas, but he was passed in Dominguillo by a courier who went up the mountain route described by Ajofrín and followed by Padre Ponce (Fig. 1). Thiéry de Menonville reported 70 crossings of the Vueltas between Dominguillo and Atlatlauca. From Atlatlauca he went up the mountains to "Galiatitlan" (Guaxolotitlán, the present-day Huitzo). He retraced his route exactly on his return.

Integration of the Cuicatec Region

I have described the economic systems of the Cuicatec towns up to now as if each stood alone. However, it is clear within the preceding descriptions that these towns were partly interdependent. Even the most remote of them was not in any real sense self-sufficient.

One model that fits the case in the Cuicatec area is the Sanders model of highland-lowland symbiosis (Sanders 1956). This symbiosis is structured, in this case, along two parameters of difference, which are linked. One is the highland-lowland contrast, with the differences in the products produced in the two different areas. The other difference is the difference between mixed irrigation and rainfall or entirely rainfall agricultural systems on the one hand, and entirely irrigated systems on the other hand. The lowland Cuicatec towns were also the towns that were entirely irrigated, so the two components were linked. However, the distinction is important because some of the symbiotic relationships between the Cañada towns and the highland towns were due to the one difference, and other relations were due to the other.

Another kind of symbiosis existed between towns with different specializations. These were less directly tied to environmental differences between the two sets of towns. The Cañada towns tended to be market towns because they were on a communication route or to be channels through which products flowed to the highlands. This is less environmentally determined than the fact highland towns produce pulque because maguey only grows in the highland zone. There is no simple environmental explanation for a town like Pápalo having become specialized in making cotton clothing, with cotton they had to go 20 leagues to get.

I will describe this kind of symbiotic relation as it exists, and try to find explanations for its existence in the system itself, rather than in any single environmental factor. This is in keeping with my attempt to see this area as a functioning ecosystem composed of subsystems, which in turn fits into a larger system.

The first, and perhaps most important interchange between the highland and the Cañada towns, was related to the greater reliability of the irrigation agriculture. The *Relación de Papaloticpac* tells that when their crops failed, the people of Pápalo bought supplies from the irrigated lands of

Cuicatlán. The Mixtec town of Almoloyas, which had several Cuicatec *sujetos*, was "vexed" by them when they withheld crops that they were able to grow with irrigation in the Cañada (probably lands belonging to Tutepetongo and Tonaltepec) to the extent that the people of Almoloyas went to Yanhuitlán to get help for the military conquest (Burgoa 1934a, I:387). Where the towns did not exchange for, or depend on, products grown by the lowland towns directly, they may have actually participated in the agricultural systems of the lowlands, depending on this in part for their subsistence. The town of Tepeucila had a pattern of migratory labor in which they rented irrigated lands in the region (presumably in the Cañada), as did the Mixtec town Xaltepetongo (*Relación de Papaloticpac* 1905:98; *Relación de Guautla* 1962:11).

Secondly, there was a kind of symbiosis such as that which Sanders describes in his initial formulation, of highland and lowland products, with exchange dictated by the sheer differences in the possibilities of the region. One of the contributions of the Cañada irrigation system was the above-mentioned secure supplies of food, of all kinds (corn, chiles, squash, beans, although beans may have been in part a highland specialty). Secondly, the Cañada farmers were specialists in fruit production.

The highlands also had their specialties. The highland towns must have traded wood for construction, just as they do today, bringing wood and charcoal down from Reyes Pápalo to Cuicatlán (Fig. 3). There are references to tribute in charcoal in Durán (1965, I:211).

> . . . de otras partes tributauan leña, cortezas de árboles, ques leña de señores por la hermosa brasa que hace, y tambien tributauan gran cantidad de carbon, y esto tributauan todos los pueblos que tenian montes: otros pueblos tributaban piedra, cal, madera de tablas y vigas para edificar sus casas y templos.

Today the Indians of Reyes Pápalo bring down vigas (roof beams), tablas (boards broad-axed from trees), and nets full of charcoal, which they trade in Cuicatlán, making one or two trips a day, and getting 10 pesos a viga, or 20 pesos for two trips. They drag the vigas by a cord through a hole cut in the end of the viga and carry the tablas and the charcoal with tumplines or breastbands (see Fig. 3 in Chapter 4). This pattern probably dates from pre-Hispanic times. Pulque and copal were important highland products. Other highland specialties would have been hunting products, and certain kinds of feathers they obtained either hunting or raising birds.

There were some products that had value that happened to occur in the highland towns. The source of the chalchihuites (green stones) that were so valued and used for tribute is not known, but they may have enriched some highland towns. Papaloticpac (Concepción Pápalo) had a small gold mine at the time of the conquest (*Relación de Papaloticpac* 1905:93). Tepeucila had a small salt spring, which was said to be barely large enough for the town itself, but which supplied salt for at least one other town (*Relación de Guautla* 1962:16).

Finally, there were a number of products that seem to have been a part of an exchange system that reached beyond the Cuicatec region. Salt, cotton, cacao, and finished work in cotton and feathers were traded over larger distances. To these, while I have no evidence from historical sources, we must add trade in obsidian, which must have been extremely important in pre-Hispanic times. The fact that it was not mentioned in the historical *relaciones* of 1579-80 probably reflects the rapid replacement of obsidian by metal with the coming of the Spaniards.

As we have seen, some of the trade in salt was within the Cuicatec area, from Tepeucila to some other Cuicatec towns. However, the salt spring at Tepeucila was described as small and diminishing. It certainly was not the source for all the salt for the entire Cuicatec area. The *Relación de Papaloticpac* (Pápalo) (1905:93) tells us that "proveese este pueblo de sal de otros comarcanos que cayn en el camino Real que va a la cibdad de Mexico. . . ." The towns "en el camino Real" were probably Cuicatlán and Teotitlán del Camino. Similarly, the *Relación de Atlatlauca y Malinaltepec* told us that

> En estos pueblos no ay salinas: proveense de sal en los "tianguez", donde se uende y la traen yndios de los pueblos de Cuzcatlan y Cuicatlan, que son ocho y catorze leguas de aqui. [1905:175]

This salt in the "tianguis" (market) in Cuicatlán came from other towns in the region:

> Proveese este pueblo de sal de los comarcanos, porque en este no ay salinas que la den, y diez leguas de aqui se cria y coje en cantidad, y en el pueblo de Tepeucila se coje alguna, avnque poca. [*Relación de Cuicatlán* 1905:188]

Most of this salt was probably from the Tehuacán Valley, traded through the tianguis, or market, at Teotitlán del Camino. While none of the Cuicatec towns specifically mentioned this town as a source, two other towns in the region—Guautla and Xocotiquipaque (*Relación de Guautla* 1962:6)—say that they got their salt from Teotitlán.

Cotton was traded through a network that reached well beyond the bounds of the Cuicatec area. Trade in cotton and cotton clothing was an important part of the highland economies, and cotton even functioned as a kind of money (*Relación de Atlatlauca y Malinaltepec* 1905:165; Hunt 1972). One of the most frequently mentioned sources of cotton was the "Rio de Albarado," which was the zone of the lower Papaloapan. For example, the *Relación de Papaloticpac* (1905:93) states that: "Para vestirse se probeen de algodon del Rio de Albarado y su comarca que . . . estara como beynte leguas poco mas o menos de este pueblo. . . ." This cotton was quite important, as,

> Tienen por contrataçion y granjeria los naturales de este pueblo conprar algodon del dicho Rio que se cria en abundançia, y esto hilan y texen sus mugeres y hazen dello naguas y mantas ques el vestido de los naturales, lo qual les vienen a conprar españoles y naturales y lo llevan a vender al propio Río donde se truxo el algodon y a otras partes, y del valor que sacan pagan sus tributos en dineros. . . . [*Relación de Papaloticpac* 1905:93]

Tepeucila also tells us that "para rropa de su vestido van a conprar el algodon al Rio de Albarado que estara como veynte y cinco leguas de este pueblo" (*Relación de Tepeucila* 1905:98). Tutepetongo continued this chain, by getting its cotton from Tepeucila: "Legua y media deste pueblo en el de Tepeuaca van los naturales á conprar sal, y el algodón para su vestir van al pueblo de Tepeucila" (*Relación de Guautla* 1962:13). This may be a mistake in the *relación*, as it seems to make more sense that they got their salt from Tepeucila, a league and a half away, and their cotton from Tepeuaca.

Cuicatlán grew some cotton, as mentioned above, and got the rest from the Río de Alvarado.

> Cojese en este pueblo algun algodon de que los naturales hazen rropa par su vestido, avnque, la mayor parte de lo que gastan es del Rio de Albarado y de otros a el comarcanos que lo trayn en yndios, questara como beynte leguas de este pueblo. [*Relación de Cuicatlán* 1905:188]

The fact that this cotton was brought by *yndios* may indicate that Cuicatlán was a market and clearinghouse for many products in the region.

Finally, Atlatlauca ranged the widest of all the towns for its cotton, going from the Río de Alvarado, on the Gulf of Mexico, to the Costa del Sur, on the Pacific: "El algodon para las mantas de que se bisten lo traen de la costa del Sur, y del rrio de Aluarado a la mar del norte, que lo vno y lo otro esta mas de treynta leguas de aqui" (*Relación de Atlatlauca y Malinaltepec* 1905:175).

Another product that was both quite important, and traded over long distance, was cacao, or chocolate. The *Relación de Atlatlauca* tells of cacao, that

> . . . hazen vna bebida de cacao, ques vna fruta a manera de almendras y corre entre ellos por moneda; muelenlo con la masa que hazen del maiz y lo deslien con agua y lo beben y le es de muy gran sustento; no todos los alcançan todas bezes, porque cuesta dinero y los macehuales no lo tienen. . . . [*Relación de Atlatlauca* 1905:171]

Cuicatlán paid tribute in cacao to their cacique. We are told that

> Comian los naturales tortillas . . . y no les era permitido beber cacao; y el Señor comia gallinas . . . y hera les permitido bever cacao y otros brevajes que no podian bever los maçehuales. [*Relación de Cuicatlán* 1905:186]

So cacao was quite important, used for money, and often limited to nobles. Even where, after the Conquest, it was not limited to nobles, it was quite expensive and out of the reach financially of commoners.

That the Cuicatecs in Cuicatlán paid some cacao in their tribute may mean that they grew some there, as they do today. However, the *Relación de Atlatlauca* (1905:175) tells us that, "El cacao se trae tanbien de la costa (del sur) avnque lo mexor es el que se trae de Guatemala y Soconusco." Thus we have a hint of a pre-Hispanic trade over much larger distances. As we know, the Cañada was the major route from Chiapas and Guatemala, so we have evidence from this material that, if cacao was traded north to the Valley of Mexico for, perhaps, obsidian, both would have passed through the Cañada.

We get a picture of the nested economic systems of the Cuicatec ecosystems. At the lowest level, each town was partly self-sufficient, with some corn cropping, as well as chiles, squash, and beans. At a higher level these towns were integrated into a larger economic system which was roughly coterminous with the boundaries of the Cuicatec ecosystem. This was based on two kinds of complementary divisions of the towns, which roughly can be described as the highland, mostly non-irrigated towns vs. the lowland Cañada irrigated towns.

The first complementary relationship is that between the certain production of the intensive irrigation system, as opposed to the vagaries and occasional failures of the highland systems which were based mostly on rainfall and also in part subject to the dangers of frost. The solution to this problem was that highland towns either rented land in the irrigated towns, or bought surplus crops in years of crop failure from the Cañada towns.

The second highland and lowland complementary relationship was that due to the environmentally-determined different possibilities of the two zones. From the highlands came such products as wood and charcoal, products of hunting, and hunted or raised birds and their feathers, chalchihuites, gold, and beans, while from the lowlands came products like corn and specialized fruits. In addition, the lowland communities were able to offer a number of products that came from far afield, because they were on the major route of communication. Thus the only markets were in the lowland Cañada communities, including possibly Cuicatlán. The markets in Teotitlán del Camino and Cuzcatlán (Coxcatlán) were out of the Cuicatec area, but they were also in the Cañada on the Camino Real.

Finally, some luxury items, and some items that were rare in their occurrence, such as salt, cacao, cotton, and presumably obsidian, caused the region and the towns to be part of a larger, sometimes even Mesoamerica-wide network. Cotton seems to have been procured directly by the highland towns. This can be understood since the source of the cotton, the Río de Alvarado, lay on the east side of the mountain system in which most of the highland towns were found, while the route up the Cañada on the Camino Real was on the western side.

However, many of the other products coming from outside the immediate region seemed to have arrived via the route later described as the Camino Real and then were carried up to the highland towns. In the case of salt this is partly due to the fact that sources of the salt lay on the Camino Real. In the case of cacao, and presumably obsidian, this would have been true because they came from far enough off that their trade was forced to follow the natural routes for travel.

Social and Political Structure

The economic system discussed above was structured and limited by the social structure by which these people in these systems were organized and through which they carried out these economic activities. Much of what we know from historical sources of the kinship, social, and political structures of the pre-Hispanic Cuicatec *cacicazgos* is from the point of view of the noble class. Aside from a few minor remarks, most of the descriptions center on the *principales* (nobles) and *caciques* (rulers). Consequently some of the following description, which is based on the *relaciones*, should perhaps be taken with a grain of salt when applied to the common people. This description will be somewhat abbreviated, as it parallels quite closely Eva Hunt's conclusions (1972), which are based on more extensive material and analysis than mine, and with which I am substantially in agreement.

The limited information available on Cuicatec kinship provides a few specifics. First, we know that they were polygamous, at least potentially. The *Relación de Atlatlauca y Malinaltepec* (1905:167) tells us that "los . . . prençipales y maçehaules tenian todas las mugeres que cada uno pudia sustentar conforme a su posible." In the case of the *cacique*, only children of his chief wife were eligible for inheritance, although the other children of other wives would be "prencipales."

> Los caçiques tenian todas las mugeres que querian, avnque entre éllas avia una que era tenida por muger natural, y solo los hijos della eredavan el cacicaçgo, y no los de las otras; y quando desta no los tenia, avnque los tuviese de otras y de las demas, no lo eredauan; heredaualo el pariente mas çercano, y este sustentaua a los hijos que quedauan del caçique, que eran tenidos como bastardos. [*Relación de Atlatlauca y Malinaltepec* 1905:167]

This statement parallels the description of the Mixtec by Herrera (1726:98) although Herrera says that when the "mancebas" who were concubines to the *cacique* had a child they would then be married off to someone else.

The nobles were class endogamous. Nobles had to marry nobles, and a *cacique* in particular had to marry the daughter of another *cacique*. Eva Hunt, has established, in part from unpublished documents, that inheritance was ambilateral, that is, that one could inherit a *cacicazgo* from one's mother, as well as from one's father. Although males were preferred as caciques, one could inherit a cacicazgo from either males or females. Hunt even mentions a female *cacica*.

Since there was only one *cacique* per town, and being class endogamous, *caciques* were town exogamous. We have no evidence as to whether this applied to a lesser extent to the other *principales* and to the commoners, or macehuales. The commoners could have been endogamous within the town. It is also not clear whether there was a difference in the legitimacy of the offspring of different wives in *principal* and *macehual* families, although the statement in the *Relación de Atlatlauca* seems to refer specifically to the *caciques*, as distinct from the rest of the population.

There is a statement that there were no dowries, but rather bride price, particularly among the *caciques*.

> No dauan dote ni cosa alguna con las hijas a los maridos, antes ellos enviauan presentes a los padres porque se las diesen, y esto avn oy se guarda entre ellos, digo entre los caçiques. . . . [*Relación de Atlatlauca y Malinaltepec* 1905:167]

In addition to the above rules about inheritance, which indicates ambilaterality with a leaning towards patrilineality, there was a further requirement for holding a *cacicazgo*. This was a residence requirement (Hunt 1972:224-25). This seems to have meant that one would have had to choose where one would be *cacique* if one inherited a *cacicazgo* in both the male and female lines, although it is possible that, like the Mixtec (Spores 1967), one could have part-time residences in each of several towns in which one was *cacique*. At any rate, even if combined in one person like this, the individual *cacicazgos* retained their separate identities and in the next generation could be inherited by separate people.

As Eva Hunt points out (1972:221), there are two basic classes in the Cuicatec *cacicazgos*: the *macehuales*, or *terrazgueros*—the common people—and the *principales*, or nobles—the upper class. Within the lower class there were slaves, who were lower than the rest, and among the upper class the *cacique* stood above the rest of the nobles. In addition there were positions within both classes that could be filled by its members that gave them some preeminence.

The bulk of the population were commoners. These are referred to in the *relaciones* by the Nahuatl-derived term "macehuales." Presumably most were engaged in agriculture. Eva Hunt mentions some evidence of specialization indicated by names, citing such specializations as salt workers, weavers, builders, painters, musicians, and house servants (1972:221). Whether these specialists were full-time, or set apart from agriculturalists, is not clear.

In the *Relación de Atlatlauca* there is a passing reference to "*esclavos*" (slaves). It is unclear whether they represented a regular part of the society. The reference says that when they needed someone for sacrifices

> sino para hazer sacrificio a sus dioses; si no avia esclauo que sacrificar escoxia el Señor el que queria y aquel auia de morir para el sacrificio y no avia mas, como dizen de tender el pescueço. [*Relación de Atlatlauca y Malinaltepec* 1905:166]

And again, "Y si en el pueblo avia esclavos, dellos matauan y sino el que el Señor escoxia" (ibid.). This may mean that esclavos were really war captives who were kept specifically until they could be sacrificed. If this is the case of all the "slaves" in the Cuicatec area, then they performed a function in the society considerably different from that which is usually understood by the term "slave."

Some Indians filled an office as *tequitlatos*. These were probably chosen from the lower class and acted as mediators between the *cacique* and the people. They directed the people to work when the *cacique* ordered something done.

> . . . declarado lo que el caçique . . . mandaua, luego se ponía en execucion, y si era negocio que los maçehuales avian de hazer, mandauase a los «tequitatos», que son vnos yndios mandones que ay en cada barrio, que tienen a su cargo los tales yndios que ay en aquel barrio para hacerlos acudir a los seruiçios que an de hazer, que comunmente aca se llaman «tequios»; estos tequitatos andan de casa en casa diziendo a los yndios lo que han de hazer, y de aquello no a de faltar ninguno, so pena de que, si era negocio que el Señor mandava, abia de morir luego por ellos aunque fuese causa bien liviana. . . . [*Relación de Atlatlauca y Malinaltepec* 1905:168]

The term *tequio*, meaning labor owed the town government, still survives. Today many people leave Reyes and Concepción Pápalo and move down to Cuicatlán, in order to avoid *tequio* labor.

The *Relación de Atlatlauca* goes on to observe that the life of the common people was extremely hard, as they had to work for the *cacique* or be punished by death: "y con esta crueldad y opresion estauan de ordinario en tanta servidumbre que casi toda uida travaxauan para los caçiques" (*Relación de Atlatlauca y Malinaltepec* 1905:168). This theme is repeated many times in the *relación*. It may reflect true greater oppression of the people of Atlatlauca than those of the other towns in the area, or an idiosyncratic feeling of the writer of the *relación*, who went into much more detail than most other writers of *relaciones*. However, it should be noted that, while all the other Cuicatec towns described in *relaciones* were under the Royal Crown ("en la Corona Real") by the time of the *relaciones*, Atlatlauca and Malinaltepec were at least half under an encomienda. The emphasis on the pre-Hispanic exploitation of the Indians of Atlatlauca and Malinaltepec may have been intended to put the treatment of the Indians by the encomenderos at the time of the *relación* in a favorable light. This is not to argue that commoners did not in fact have fairly onerous lives.

The upper class had one clear-cut distinction, that between the *cacique* and the rest of the *principales*. Membership was defined by birth, and birth defined the *cacique*. The distinction between the *cacique* and the rest of the *principales* was not quite absolute, as there were lesser *caciques* who ruled barrios or *sujeto* towns (Hunt 1972:218-19). Another position drawn from the *principales* was that of the elders who counseled the *cacique*. One of these in particular worked as a kind of mayordomo and executive secretary and chose which cases could come before the *cacique*. This official had a separate patio and reception room and received complaints from the people, as well as embassies from other towns. He communicated with the *cacique*. When the *cacique* made a decision the old man was the one who announced it. The decision was then put into operation by the *tequitlatos*.

Also drawn from the noble class were the priests. These were trained from childhood. It seems the *caciques* always went through this training as well. From those trained the *cacique* chose the priests. They served for terms of seven years. They were said to have been more powerful than the *cacique*, but since they were chosen by the *cacique* and their term limited, this statement must be taken with a grain of salt.

Presumably the *principales* were expected to be leaders in time of war. When the *cacique* did not lead the army, he appointed a *capitán general*. However, the *capitanes* were chosen for bravery, so it is possible that commoners could rise through the wars. It is also unclear if there was a way that *principales* could drop far enough in the social scale to become commoners, as they became more and more removed from the main line of descent. This sort of question is intriguing, and someone more adept at ethnohistory than I should examine it.

The noble class was distinguished from the commoners in several ways. The *cacique* was treated "like a god." People had to take off their shoes before entering his presence and had to lower their eyes when in his presence. The rest of the nobles were distinguished by dress. "Los Caciques y Señores sienpre se aventajaron en el tratamiento de sus personas y bestidos" (*Relación de Papaloticpac* 1905:91). "De hordinario andavan desnudos, con solo vn paño . . . y los señores trayan el propio paño y mantas rricas" (*Relación de Tepeucila* 1905:96).

> . . . estas mantas eran listadas de colores y texidas muchas labores por abaxo. Tenian vna como çanefa hecha de labores y entretexida por ellas plumas blancas y otras colores. . . . Estas mantas traian los prençipales, y las de los maçehuales eran de nequen, ques vn hilo que sacan de las pencas de maguey, y del hazen vna tela muy grosera, y aun muchos dellos avn esto no alcançauan, y andavan en carnes con . . . solos pañetes de nequen . . . los saçerdotes traian estas mantas çeñidas al pescueço con un cordel y en esto se conoçían. Traian orexeras y beçotes de oro los caçiques y los prencipales, y quentas al pescueço de chalchihuites y de oro. . . . [*Relación de Atlatlauca y Malinaltepec* 1905:170]

The *principales* were also distinguished in having certain exclusive foods and drinks (see discussion above), especially cacao, deer, and possibly turkeys, and all hunted meat.

The minimal unit in which a Cuicatec society in its complete form could be found was a single town. In these towns one could find the *macehuales*, the *principales*, and the *cacique*. The town might be divided into *barrios* (contiguous or nearly contiguous neighborhoods within the town) or the town might have other towns which were *sujetos* to it. A *sujeto* would be another, physically-separated population settlement, which was subject to the government of the *cabecera* (head town) and which contributed to the *cabecera*. These smaller subdivisions might simply be governed through a (commoner) *tequitlato*, or it might be under a *principal* called a *cacique*, but who was under and acknowledged the superiority of the *cacique* of the *cabecera*.

Within this minimal unit, the *cacicazgo*, orders flowed down from the *cacique* through his mayordomo, to the minor *cacique* or the *tequitlato*, who directed the actual work. In turn, the *cacique* was supported by the tribute which the people had to give to support his household. The *Relación de Cuicatlán* distinguishes this tribute clearly from that due Montezuma:

> . . . era Señor de este pueblo Tiñaña, que tenía este nombre en su lengua propia . . . al qual tributavan los naturales que tenia por sujetos, mantas, cacao, maíz, y frisoles y axi y aves y pescados y todas las demas cosas que avia neçesidad para su sustento y vestido. [*Relación de Cuicatlán* 1905:185]

This tribute seems quite homely and down to earth when contrasted with the rare items in the tribute to Montezuma: "y para quel dicho su Señor tributase a Montecçuma, de quien era su sujeto, buscaban los naturales plumas, oro, piedras" (ibid.).

Even including the *sujetos* and separate *barrios*, and making liberal allowance for population decline from the time of Conquest, the largest of these units was considerably under 5000 people in population, and all the rest under 3000 (these figures were arrived at by rounding off the highest reasonable sixteenth-century population estimate, as derived in Table 3, upward to the nearest thousand). Even if we add a few thousand to this unit for a further margin of error, it is clear that the individual units were really not very large.

The fact that, while apparently truly stratified, the Cuicatec towns were of such a small order of size, presents a problem in terminology. This terminology problem is important because the terms imply important assumptions about the nature of the culture.

One hesitates to apply the terms "city," "state," "city-state," and "civilization" to a settlement of 3000 people. Sanders and Price, for example, make the following points about minimum population and the terms "town," "city," and "civilization."

> The size of the population is functionally related to the degree of internal differentiation. We would, therefore, make a distinction between towns, urban communities with population in the thousands, and cities in the tens of thousands. [Sanders and Price 1968:46]

> Civilizations by definition are large social systems characterized by intense social stratification and economic specialization.... Our estimate is that such systems require a minimum of 10,000 individuals to function. [ibid.:74]

Eva Hunt has recognized that this is a problem, and has struggled gamely with it. She uses the term *cacicazgo* in the title of her article. She refers to the Cuicatec "state" (1972:171), "minor 'kingdoms' or states" (1972:211), throughout the work. She discusses this terminological problem, pointing out that the *cacicazgos* are stratified, but quite small. She suggests that some kind of term like "village state," "town states," and "micro-states" might be better (1972:237-38). She even mentions the possible applicability of Ronald Spores' term "community kingdom" (ibid.). Finally, she suggests that the "micro-states" of the Cuicatec area may not have really reached what Durkheim called "the dynamic density" for the complex division of labor essential to a complex urban state (1972:241).

Since this study addresses both the *level* of political integration reached by the Cuicatec towns and *how* this level was reached, the terminology is important. This chapter is largely descriptive, laying out the information available. I would prefer to present the issue, but not prejudge the question by picking a term that commits me to an opinion. For that reason I will use the term *cacicazgo* for this minimal unit of the Cuicatec population.

A *cacicazgo* is a little community, in Redfield's terms (1971). It is the smallest complete microcosm on which all, or almost all, of a culture can be described in all of its workings. A *cacicazgo* is the entire geographical and political unit under one *cacique's* rule, whether this is one single population settlement, with or without *barrios* as subdivisions, or several population nuclei arranged in a hierarchical *sujeto-cabecera* relationship.

It remains to be seen whether these *cacicazgos* were combined in some larger political unit, which might more satisfactorily and unequivocally be called a state. Burgoa describes a dispute that the town of Almoloyas had with some Cuicatec "estanzuelas." Almoloyas was a Mixtec town, but some of its *sujetos* were Cuicatec (Tonaltepec and Tutepetongo?). He says of these estancias, "Cuicatecos, de estos eran muchos y con estar repartidos por la sierra del Oriente, tenian de su nación inmediata una provincia grande, con quienes se comunicaba . . ." (1934a, I:387). By *nación* Burgoa apparently seems to mean people of the same linguistic group, although he states that these people did communicate and hence presumably felt some kind of unity. By *provincia* he seems to imply the existence of a larger Cuicatec political unit of some sort. However, in the military dispute that he describes, in which the Mixtec of Almoloyas went to the Mixtec of Yanhuitlán for help against the Cuicatec, it is clear from the description that the Cuicatec who were conquered were those of the "estanzuelas" subject to Almoloyas, and not all the people of the Cuicatec "provincia."

I have discussed above the economic ties that tend to draw the Cuicatec towns together as a unit of some sort. The symbiotic relations between the highland and lowland towns, the exchange of special products, such as salt, cotton, and cacao, and the existence of a market and possibly itinerant merchants at Cuicatlán, all would tend to create at least some cohesion between the Cuicatec towns.

Eva Hunt states that there is no documentary evidence that supports the claim of the present-day inhabitants of Cuicatlán that their town was the capital of a pre-Hispanic pan-Cuicatec "state." However, she points out several other factors that reinforce an integration among the Cuicatec-speaking *cacicazgos*.

First, the rule of class endogamy and the rules of

inheritance of *cacicazgos*, would mean that the *caciques* would of necessity be town exogamous (Hunt 1972:206). This would mean that *caciques* would be related to other *caciques* throughout the region, creating ties that reach beyond the individual *cacicazgos*. This was reinforced by a mythology of common descent of all the Cuicatec *caciques* (Hunt 1972:206). In addition, she argues that there was a religious and ceremonial cycle of the various gods of the pantheon divided among the towns, which would induce an interdependence based on religion that paralleled and reinforced the economic interdependence and symbiosis.

Finally, Hunt points out that, "All . . . cases of conflict involving named towns, settlements, or states involved warfare with non-Cuicatec-speaking towns" (1972:206). Because of this she argues that

> This strongly suggests that the Cuicatec town-states as a group were more than just speakers of the same language, but were economically and politically allied, while their enemies were speakers of other languages and attached to cacicazgos of other ethnic groups. This interpretation is supported by our evidence on internal trade and cacique intermarriages within the Cuicatec state, and suggests that the solution of internal territorial conflicts between Cuicatec states was by peaceful means, and that the political elite of the Cuicatec formed a single macro-descent group. [Hunt 1972:212]

However, while pointing out that there is, as stated above, some sense in which all the Cuicatec towns did stand together,

> There is no evidence . . . that the Cuicatec had a formal "confederacy," or that all the Cuicatec states were unified under a single ruler. Kinship loyalties of the ruling class and intra-cacicazgo loyalties following kinship links of the nobility promoted fusion. But the same served to promote friction. . . . In this sense, 'royal kinship' as a basis for the unification of territories, was as poor an instrument of politics as it was among nineteenth-century Europeans. [ibid.:206]

The statements we have on warfare in the *relaciones* help us deal with this question. The *relaciones*, it should be remembered, not only talk about warfare that preceded the some 60 years of Pax Hispánica, but also a certain number of years of Pax Azteca preceding the Spanish Conquest. We are told that before they were conquered by the Aztec armies, various towns in the Cuicatec area warred with each other. Cuicatlán, before it was conquered, had "guerra ordinaria" with the lord of Quiotepec. The two highland towns of Papaloticpac (Pápalo) and Tepeucila are said to have fought with all the other towns in the region. "Tenian ordinaria guerra con todos los pueblos comarcanos" (*Relación de Papaloticpac* 1905:90). "De ordinaria tenian guerra con todos los pueblos de su comarca y ellos con los de este pueblo, solo a fin de cativarse para sus sacrificios" (*Relación de Tepeucila* 1905:96). These statements do not specifically say that Cuicatec towns fought other Cuicatec towns, since all had non-Cuicatec towns in their region (*comarca*).

With only four towns this difference is probably not significant; however, it is interesting that in the *relaciones* of the two highland towns of Papaloticpac (Pápalo) and Tepeucila they are said to have fought their wars to capture slaves for sacrifices, while in the lowland towns of Atlatlauca and Cuicatlán they fought to capture lands and tributaries, as well as captives: "el premio de la gverra era hazerse esclauos de los vençedores, y otras vezes quedan por tributarios dellos, . . . como lo fueron de Monteçuma" (*Relación de Atlatlauca y Malinaltepec* 1905:169).

> . . . antes que muriese el Tecuantecotle y lo sujetase Monteçuma tenia guerra ordinaria con otro Señor del pueblo de Quiyotepeque que se dezia Tico en su lengua y . . . guerreavan los vnos con los otros, por quitales sus tierras, haciendas, y por cativallos para sacrifiçios. . . . [*Relación de Cuicatlán* 1905:186]

There was a plot of land on the Río Cacahuatal between Cuicatlán and Quiotepec which could have been controlled by either of the two *cacicazgos*. We are informed that they ate people killed in war, but this was probably not a major reason for war.

If there was inter-Cuicatec warfare, it would argue against a regional Cuicatec unity. There is evidence that at least some of the relationships of a political nature cut across linguistic divisions, especially on the borders. Atlatlauca was Cuicatec and Malinaltepec Chinantec. This is probably the source of Herrera's statement that Atlatlauca was Chinantec, since it was part of a political unit with dual linguistic affiliation. Similarly, Eva Hunt (1972) cites evidence that Quiotepec, which was mixed, but probably mostly Mazatec, may have at one time paid tribute to Cuicatlán. Some small Cuicatec towns (probably Tonaltepec and Tutepetongo) were conquered by the Mixtec of Almoloyas, with the help of other Mixtec from Yanhuitlán. And Teutila, which seems to have been Cuicatec, was *cabecera*, at least at a later time, of an area including many *sujetos* who were either Mazatec or Chinantec (Ajofrín 1959).

In summary, then, I agree with Eva Hunt that there was a definite larger unity of some kind between the Cuicatec *cacicazgos* on the basis of kinship between the *caciques* and linguistic affinity. This was reinforced and paralleled by symbiotic relations between these *cacicazgos*. However, as Eva has pointed out, there were considerable forces for schismogenesis. Relations were by no means limited to linguistic lines, and loose ties also existed across linguistic lines. In no way could this unity be considered a "confederacy," or "state," or any kind of effective political unit under one central government, even of the weakest sort. The Cuicatec *cacicazgos* were like the Mixteca Alta:

> No substantial evidence indicating the existence of a supreme Mixtec monarch with broad authority over an extensive Mixtec-speaking domain had yet come to light. . . . In all of the sixteenth-century documentation, we find little to preclude the existence, on the eve of Conquest, of a number of separate and autonomous kingdoms. There is evidence that two or more of these kingdoms were sometimes temporarily united under a single ruler (*señor*). But these multiple estates could have also been divided among the ruler's children. [Spores 1967:67]

This whole analysis of the size of the largest political integration of the Cuicatec *cacicazgos* is complicated by the fact that the Cuicatec *cacicazgos* had been part of the Aztec empire for a number of years before the Conquest. If a larger political unity among the Cuicatecs existed, it was submerged by the Aztec conquest.

Atlatlauca and Malinaltepec say they were conquered between 10 and 20 years before the Spaniards came although nobody could remember exactly (*Relación de Atlatlauca y Malinaltepec* 1905:165). Tutepetongo does not say when it was conquered, but its *señor* at the time of the Conquest was a *capitán de guerra* for Montezuma. Tonaltepec was tributary to Tilantongo, in the Mixteca Alta, at the time of the Conquest. Tepeucila says that they were conquered by Montezuma during the lifetime of the individual who was *cacique* when the Spaniards came.

> Al tiempo que vinieron los españoles era Señor de este pueblo vn cazique que se dezia Canchuchu Camiñaa, y en tiempo de este los sujeto Monteçuma . . . (*Relación de Tepeucila* 1905:95]

Cuicatlán also says that it was conquered by Montezuma, although at the time of a *cacique* who died before the Spaniards came (*Relación de Cuicatlán* 1905:185-6).

Barlow includes most of the Cuicatec Cañada towns in the province of Coayxtlahuacán. Barlow justified his identification of the "provinces" on the basis of inclusion in groups in the *Matrícula de Tributos* (Barlow 1949). This he defends by stating that they are included on the same page and turn out to be geographically contiguous. Further, he argues that there is linguistic unity in many of the groups and that they often represent places known to have been conquered in one continuous campaign by the Aztec armies.

If, as Barlow suggests, the Cañada towns were conquered as part of the province of Coayxtlahuacán, then they would have been conquered in the first subjugation of that town, by the first Montezuma, as reported by Durán (1965). Coixtlahuaca later rebelled and had to be reconquered. Ixtlilxochitl, in a later, more condensed, and more Texcoco-oriented version of the same events (Ixtlilxochitl 1965), attributes this conquest to Texcoco in 1486, giving Nezahualpiltzintli, the Texcocan head in the Triple Alliance, all the credit, although later, when Nezahualpiltzintli was older and Coixtlahuaca rebeled, Ixtlilxochitl gives Montezuma credit for reconquering it (Ixtlilxochitl 1965:318), when sent by the head of Texcoco.

This period of the reign of Montezuma the First was the period in which the empire was being extended through the Mixteca. If the Cuicatec were not affected by the conquest of Coixtlahuaca, there are several other campaigns that would have brought the Aztec armies through the area, as one of the two natural routes south through the highlands. The first of these, after the fall of Coixtlahuaca is the conquest of Oaxaca (apparently the Valley of Oaxaca) in which the city was completely destroyed. Later were the campaigns against Tehuantepec. Burgoa, referring to one or the other of these campaigns against the Zapotec, states that there was a major frontier fortress of the Zapotec at Guaxolotitlán (the present Huitzo). The Mexican armies, not being able to come through the Mixteca since the Mixtecs were in rebellion, came through the Cañada to attack this fortress.

> . . . del mexicano que entraba talando y abrasando las poblaciones que sujetaba, y no pudiendo abrir paso por la mixteca, le hizo por el camino real de los cuicatecos, y éstos por la parte del Norte llegaban a los umbrales de esta frontera del zapoteco que se llamó en su lengua *Huijazoo* que significa atalaya de guerra. . . . [Burgoa 1934a, II:11]

It is not clear whether the Cuicatec were previously subjugated or not, or whether they were conquered or simply gave up when the Aztec armies showed up. At any rate their subjugation undoubtedly dates at the latest to this series of campaigns.

Some of the Cuicatec towns were included with Mixtec towns. The town of Almoloyas was formed of a number of estancias, two or three of which were Mixtec, the rest being Cuicatec. Before the Mixtec were conquered by the Aztec, the Cuicatec in Almoloyas, who controlled lowland irrigated fields, began to refuse to give part of their crops to the Mixtec. The Almoloyas Mixtec went to Yanhuitlán to get help to reconquer them. These Cuicatec were said to be in contact with a large province of Cuicatec. This would have been at the southern limits of the Cañada, on the Río de las Vueltas above Dominguillo. We know that Tonaltepec and Tutepetongo were subjects of Guautla (not Huautla Jiménez), which is Mixtec, and which was part of the same province as Almoloyas, so these two Cuicatec towns may well be the ones conquered by Almoloyas, with the help of Yanhuitlán. This is supported by Gerhard (1972), who includes these two under the Mixtec state of Yancuitlán (Yanhuitlán). They would then have fallen to the Aztecs through the falling of Yanhuitlán and Coixtlahuaca. The *Relación de Guautla* tells us that Guautla (the *cabecera* to which Tonaltepec and Tutepetongo were subject) paid tribute to Cuestlauaca (Coixtlahuaca) which we know from the *Relación de Atlatlauca y Malinaltepec* (1905:165) was an Aztec garrison, and, according to Barlow (1949), was the head of the province to which all the Cuicatec towns paid tribute. The *Relación de Guautla* tells us that the señor of Tutepetongo was a *capitán* of Montezuma, and that they paid tribute to him, so their tribute may have ended with him. Tanatepeque (Tonaltepec) was said to have been subject to a *señor* from Tilantongo (also Mixtec), and they paid him no tribute except the obligation to go to war for him. So the relation with Guautla may represent a post-Conquest hierarchy, but there is no question that these towns pre-Hispanically were subject to Mixtec towns (= "kingdom"). Presumably most of the other Cuicatec towns were conquered directly by the Aztec, but these two towns would have fallen with their Mixtec *cabeceras*.

The situation under Aztec rule was attested in the *relaciones*, with the Pax Azteca declared. While the campaign against Tlaxiaco for attacking the tribute carriers of Coixtlahuaca probably did not affect the Cañada, it reflects the new situation in which the Mixtec of Coixtlahuaca looked to the Aztec for support in the war against Tlaxiaco.

The Cuicatec in the Aztec Empire

Whatever arguments we make about the level of and size of pre-Hispanic Cuicatec political units, we must recognize that they were not independent polities at any time within the immediate period of the historical sources we have. At the time our earliest historical sources were recorded, the Cuicatec *cacicazgos* had been conquered by the Aztec for at least a generation. Therefore we must consider how the Cuicatec *cacicazgos* functioned as part of the macrosystem of Mesoamerica. The most important aspect of this was the Aztec "empire" (for lack of another, commonly accepted term).

This does not mean that all the preceding discussion of the political units of the Cuicatec is meaningless. There is considerable evidence that the Aztec, like the later Spanish, system left the local political structure intact to a large extent. The original *caciques* were usually left as rulers of their people, and the people directly owed fealty, not to the Aztec "king," but to their cacique, and he to the Aztec ruler. Many of the *relaciones* contain statements that the people had only acknowledged one "*señor*," even though in other parts of the *relaciones* they state unequivocally that they were conquered by and paid tribute to the Aztec rulers:

> Solo el Señor natural era obedescido y rrespetado de los naturales y el los gobernava y mandava, y avnque vinieron en sujeçion de Montecçuma y de sus Capitanes. . . . [*Relación de Papaloticpac* 1905:90]

> Al tiempo que vinieron los españoles era Señor natural de este pueblo vn cazique que se dezia Canchuchu Camiñaa, y en tienpo de este los sujeto Monteçuma. . . . [*Relación de Tepeucila* 1905:95]

> [They paid tribute to the Aztecs and had to send warriors when asked.] En lo demas, ni Munteçuma ni sus gentes, no se entremetian, antes dexauan el mando y el gouierno a los caçiques y Señores naturales que en cada pueblo tenían. . . . [*Relación de Atlatlauca y Malinaltepec* 1905:166]

> A solo Tecuantecotle obediçian los naturales de este pueblo, sin rreconocer sujeçión a otro señor, avnque, como esta dicho, el enbiava en manera de tributos o parias a Montecçuma el tributo declarado. . . . [*Relación de Cuicatlán* 1905:185-86]

A little different from these are the towns of Tutepetongo and Tonaltepec, both of which were *sujetos* at the time of the Conquest, although both were Cuicatec. One was subject to the *señor* of

Tilantongo, in the Mixteca Alta. We are not told if there was a local sub-*cacique* or other single town head in Tonaltepec. Tutepetongo had as a *señor natural* a captain in Montezuma's army. They paid him tribute, rather than to the Aztec, and went with him to war when he had to participate in campaigns. This leads one to wonder if the *señor natural* of Tutepetongo was not perhaps recalcitrant and was replaced directly by a Mexican. Or perhaps Tutepetongo, as an estancia, was granted in fief to a captain. The final possibility is that the Cuicatec *señor natural* became a captain in the army of Montezuma, and as such his town was spared any tribute other than maintenance of their lord and the obligation to fight for him, when called upon by the Aztec ruler. I cannot resolve these various possibilities; I simply mention them because this one *relación* stands out from the pattern of the others and suggests one part of the process of incorporation of towns into the Aztec "empire," which is a contrast to the reports from the other Cuicatec towns for which we have information.

The local *señor natural* was supported by the *cacicazgo* of which he was *cacique*. In general the tribute for him is listed separately from that due Montezuma. This suggests that the tribute due the *cacique* represents a pre-Aztec system which was still, to a certain extent, autonomous from the Aztec system of tribute collection.

The tribute paid the local *señor natural* was essentially his food, clothing, and the maintenance of his establishment, including right to service by rotating crews of commoners. Thus the *Relación de Cuicatlán* (1905:185) tells us that

> . . . era señor de este pueblo Tiñaña, . . . al qual tributavan los naturales que tenía por sujetos, mantas, cacao, majiz, y frisoles y axi y aves y pescados y todas las demas cosas que avia necesidad para su sustento, y vestido.

The *relaciones* of Pápalo and Tepeucila, on the other hand, do not specify that their *cacique* received other tribute than that they paid to the Aztec, or if they simply gave him a cut.

The *Relación de Tutepetongo* says,

> En tiempo de su gentilidad tenían por Señor á Avaslaslac que quiere decir culebra resplandeciente, é como a su Señor natural le hacían sementeras y le servían con comidas, y le tributaban mantas de algodón pintadas de figuras de leones é águilas. [*Relación de Guautla* 1962:12]

Eva Hunt (personal communication) suggested that these mantas of "leones" and eagles could be jaguar and eagle uniforms for the Aztec military orders, since the Spaniards sometimes call jaguars "leones." If this was so, the lord of Tutepetongo, who was a captain in Montezuma's army, probably belonged to one of these orders.

Each *cacicazgo* also had to pay tribute to the Aztec empire in some form; these items were given to Aztec tribute collectors. The *relaciones* point out what people did to amass the tribute. This was an important part of the economy, for earning tribute was given as the reason the inhabitants of the highland towns rented lands in the Cañada and worked on the irrigated lands of the Cañada as laborers.

We can identify the tribute paid by the Cuicatec towns from two major sources. One is the *Códice Mendocino*, and the other is the part of the *relaciones* which describes the tribute paid to Montezuma before the Spaniards arrived. This tribute structure has been ably analyzed by Barlow (1949). As has been mentioned earlier, Barlow includes the Cuicatec towns in the province of Coayxtlahuacán (Coixtlahuaca).

The province of Coixtlahuaca, as it is identified by Barlow, probably, indeed, almost certainly represents an Aztec administrative unit, rather than any pre-existing political entity. The importance of Coixtlahuaca to this "province" is that in it was located an Aztec garrison, and presumably an administrative center.

The kind of tribute paid by the Cuicatec towns is interesting. The Cuicatec area was 350 km from the Valley of Mexico. At a time when the only means of freight transportation was the human back, this represented 1 to 2 weeks of travel. This imposed economic limitations on the kinds of tribute that would be worth transporting that far. The Cuicatec area was also far enough from the core of the Aztec empire that direct administration from the Valley of Mexico would have been difficult.

The Cuicatec tribute reflects these two factors. While the tribute to the *cacique*, where it is mentioned in the *relaciones*, was frequently in perishables such as foodstuffs for his needs, the tribute paid to the Aztec empire falls into two main categories, depending on whether it was destined for the Aztec capital or for the nearby regional garrisons.

Some of the tribute was in the form of rare or valuable items that represented a combination of high value and low bulk, which could be economically transported to the center of the Aztec system in the Valley of Mexico. Such things as gold, feathers, skins of animals, chalchihuites, and finished products in cloth and featherwork fall into this category, and almost invariably are said to have gone to Montezuma. Thus Pápalo sent gold to Montezuma (*Relación de Papaloticpac* 1905:90), Tepeucila gold, feathers, "Mantas y cueros de tigres" (*Relación de Tepeucila* 1905:95). The *Relación de Atlatlauca* is unusually complete in its statement about tribute:

> ... Muntecuma era el Señor universal de todos, y el Señorio que sobre ellos tenia era çierto tributo que le dauan en cada pueblo de çierta cantidad de grana cochinilla, mantas de algodon y plumeria verde, y de todas colores, y unas piedras verdes que aca llaman «chalchihuites» que son de poco preçio.... Para recoxer este tributo venian dos yndios prençipales de parte de Muntecuma, que se llamava «calpisques», y estos lo hacían rrecoxer y lo lleuauan a la prouinçia de Cuestlauaca, donde el dicho Muntecuma tenia puesta su frontera de gente de guerra.... [*Relación de Atlatlauca y Malinaltepec* 1905:165]

The *Relación de Cuicatlán* also mentions these officials who collected the tribute:

> Y para quel dicho su Señor tributase a Montecçuma, de quien era su sujeto, buscaban los naturales plumas, oro, piedras, y a los cobradores que venían a cobrar este tributo por Montecçuma, les davan mantas, comida y otros presentos, sin lo que pagavan al gran Señor que hera Montecçuma.... [*Relación de Cuicatlán* 1905:185]

This last suggested that in addition to paying the tribute, the towns also paid the living expenses for the employees who collected it. That is, the conquered towns paid for their own administrative expenses. Barlow (1949) and Eva Hunt (1972) interpret this tribute in food and clothing that was given the *calpixque* as bribes. The bribes may have been a necessary part of the operation of the system (as it is in many political systems today that depend in part on institutionalized bribery to pay their public servants). However, since the practice was recorded with the rest of the tribute in the Spanish *relación*, it seems quite possible that this was in fact intended to be pay for the *calpixque*. Whether bribes or pay, these forms of tribute seem to have been a necessary part of the functioning of the Aztec empire.

Another kind of tribute contributed to the maintenance of the local Aztec forces, especially those in the regional garrisons. Thus, in addition to the precious tribute, Pápalo sent "bastimentos que llevavan a las guarniçiones donde Montecçuma tenia gente de guerra" (*Relación de Papaloticpac* 1905:90). Similarly, Cuicatlán also "algunas vezes enbiavan frutas a los questaban en las guarniçiones de Montecçuma en la provinçia de la Misteca" (*Relación de Cuicatlán* 1905:185).

Some of the towns contributed even more directly to the Aztec forces controlling the region, by having an obligation to supply soldiers whenever the Aztec army went to war. Atlatlauca, in addition to the above-mentioned tribute in precious materials, "quando los capitanes de Muntecuma les mandavan que embiasen gente de guerra para ir a otras conquistas, lo hacian" (*Relación de Atlatlauca y Malinaltepec* 1905:165-66). I have already mentioned that the *cacique* of Tutepetongo was a captain of Montezuma, and the primary obligation of this town was to maintain him and serve as his warriors when he went to war. Tonaltepec was tributary to the town of Tilantongo, in the Mixteca Alta, and its only tribute was the obligation of helping him in war.

> Antes que los españoles vieniesen, tenían en este pueblo por Señor natural a un indio que se decía Ixtetecoani, que quiere decir uña de gran león el cual residía fuera de este pueblo en el pueblo de Tilantongo, en esta Misteca alta, al cual no le daban ni tributaban cosa alguna, más de que cuando los llamaba para la guerra contra otros pueblos iban con sus arcos y flechas é con las armas que tenían a le ayudar. [*Relación de Guautla* 1962:15]

If Tilantongo was conquered by the Aztec, this meant that there was a chain through Tilantongo to Tonaltepec, and if it remained unconquered, then they contributed Cuicatec soldiers who could well have found themselves fighting other Cuicatecs on the Aztec side, in the Aztec wars of conquest.

Thus the tribute of the Cañada throws an interesting light into how the Aztec "empire" was able to integrate the various different polities and linguistic groups. The crucial problem was that transportation in pre-Hispanic times was made difficult by two factors—the lack of any draft animal other than man, and the rough nature of the Mesoamerican terrain. As a result, the cost of transportation over large distances was relatively high, making the effective distance for which it was worth transporting products fairly limited. The population that could be drawn on for the production of foodstuffs that could be carried directly to Tenochtitlán-Tlatelolco, was limited to the area around the Valley of Mexico.

The same limitation of transportation meant that there was also an effective limit beyond which areas could not be directly administered from the Valley of Mexico. For this to happen, troops would be forced to travel for several weeks to get to where they would have to fight, and then when they got there, they would have had to treat every war as a re-conquest, without local lines of supply. This would effectively have limited the sphere of influence to the area which could be supplied and conquered from the Valley of Mexico.

Cuicatlán and the Cuicatec area were outside the area that could have served as a base for the food supply for the Aztec city, as the transportation costs would have made it prohibitive. They also probably were beyond the reachable limits for permanent control by a government relying on troops based in Mexico City. That they were in fact part of the Aztec "empire" is the solution developed by the Aztec system. This solution was to establish secondary centers, in which Aztec garrisons were stationed, that served as centers for the amassing of tribute and the control of the countryside. In the case of the Cuicatec area, the relevant center was at Coixtlahuaca, in the Mixteca Alta.

To this center the surplus food production was taken for support of the garrison. In this way the most common kind of surplus could be mobilized by the central Aztec government, without prohibitive costs for transportation. Another kind of "tribute" served the same purpose that food for Aztec soldiers did—that towns either fed Aztec troops when they passed through on a campaign or supplied troops for the campaign, when needed. In one case (Tonaltepec), it seems that the *cacique* of the town was a *capitán de guerra* and either was a *capitán* because his town supplied troops (and nothing else), or was awarded the town because he was a *capitán de guerra*.

While this system allowed the Aztec "empire" to extend itself beyond the area that could be controlled directly from the Valley of Mexico, it remains to be seen why they would get any profit from extending that far. The answer to this lies in the existence of certain luxury products that represented enough concentrated value to justify the transportation. This was represented by tribute like chalchihuites, cacao, gold, skins of animals, and finished products like featherwork, that were transported to the center in Tenochtitlán, where they could be amassed and redistributed if necessary. While it is outside the scope of this book, the relationship of tribute to economic factors governing it, especially transportation, is worth a more thorough analysis.

We get an idea of the functioning of the larger, almost Mesoamerica-wide system of the Aztec "empire," of which the Cuicatec region was but a small piece, from the information and description we have of the smaller Cuicatec region-wide system. In order to understand this Cuicatec region as a functioning system we must see it as a piece in its larger context. Try as we will to reconstruct how and what the Cuicatec political units were like before the conquest by the Aztecs, we must recognize that all our historical sources really describe the Cuicatec towns as they were well after the Aztec conquest (not to mention influences from the later Spanish Conquest and the Spaniards' attempts at regularization after their conquest). One wonders, after all, what was the return to the tributaries of the Aztec "empire" in return for all the tribute and service they contributed. We have, after all, evidence for some sort of Mesoamerica-wide integration at several periods. One would expect that there would be some benefit flowing from the macro- to the micro-systems, if the macro-systems keep cropping up.

The first, most obvious, benefit of being part of the "empire" was the negative benefit of not being destroyed by the Aztec armies. The Cañada towns, at least, could not avoid being overrun, since they lay on a vital north-south route, along which the Aztec armies had to travel to control other areas, such as the Zapotec and Mixtec area, and especially the Valley of Oaxaca.

However, once conquered, and as part of the Aztec system, they did benefit from the Pax Azteca. The *relaciones* state several times that before the Aztec armies conquered each town, they used to war among themselves, but that after the conquest Montezuma had forbidden it. There are examples of Mixtec people, who, after a few decades of Aztec rule, when attacked by other Mixtec groups, appealed to the Aztec armies rather than having to take their own chances. Aside from the advantage of not always having to fight one's own battles and indeed, due to a deterrent effect, frequently not having the situation develop that anyone will have to fight, one has the added advantage that when one is involved in a war, with the help of the massive Aztec support one can be sure than when one is attacked, one will emerge on the winning side.

An obvious advantage of the Pax would be allowing a greater level of regional and interregional interchange. Without a relatively stable situation, and control of war, one wonders if Cuicatlán would have been able to trade its fruit to the Valley of Oaxaca, cutting across several political and linguistic units, or if the various highland towns would have been able to engage in the complex long-distance interchange of cotton and processed cotton goods that was important to them.

To carry this to a larger context, one can argue that the larger the region effectively brought under one system, the greater the advantage. This is because the larger the area encompassed in the system, the more that local disasters can be minimized, since other areas included in the system may be unaffected and can alleviate the first area's emergency situation. On the Cuicatec regional level we have seen this kind of interaction between Pápalo and Cuicatlán. When this interaction is enlarged to the scale of the Aztec empire, the whole effect is magnified.

At any rate, the Cuicatec *cacicazgos*, as described in the sixteenth-century sources, were definitely part of the Aztec "empire." Consequently, whatever the largest Cuicatec political unit may have been before the Aztec conquest, at the time of contact they stood as individual towns, or small town *cabecera-sujeto* clusters. While there seemed to be a certain feeling of the Cuicatec region, the towns, and apparently their *caciques*, were by no means limited by linguistic boundaries in forming their political alliances. The most important fact about these alliances, however, seems to have been that they were never very far-reaching.

Similarly, no matter how autocratic the Cuicatec *caciques* may have been, they were, at the time of contact descriptions, limited to being local "lords" whose power did not extend beyond their town and who held their right to control the commoners in their towns solely in exchange for passing a share of their surplus on to the Aztec lords. This could be in the form of goods, or services in war. One might argue that they were once mighty lords, reduced in power by the imposition of the Aztec authority, but a counter-hypothesis might be offered: that they were in fact puffed up in their authority by the Aztec system, in imitation of the Aztec hierarchy.

This is important when considering the evidence of stratification in the Cuicatec *cacicazgos*. There is considerable evidence for at least a two-class society among the Cuicatec *cacicazgos* at the time of the Conquest, and the *Relación de Atlatlauca* describes (with some indignation) a system of considerable autocratic power for the *cacique* and *principales*.

But before one accepts entirely the premise that these were independently evolved, fully-stratified societies, that were later incorporated into a larger, more complex system, one should also recognize that the Cuicatec towns lay on a major north-south route, in contact with long pre-existing stratified systems in the Valleys of Oaxaca and Tehuacán, and beyond. It is possible that the Cuicatec stratified society evolved precociously in imitation of the richer centers to the north and the south. One wonders if the Cuicatec *cacicazgos* might fit Leach's *gumsa-gumlao* model from highland Burma (Leach 1964).

Leach describes the case of the Kachin, of highland Burma, whose slash-and-burn system neither demands, nor really is capable of supporting a fully stratified system. They are in contact with lowland Shan states, which are based on irrigation systems. These Shan states have a high status relative to the Kachin villages. As a result the Kachin tend to imitate the sedentary stratified Shan pattern, developing hierarchies in their villages.

In the words of Pedro Armillas (in "the Human Career," introductory graduate course in archaeology and physical anthropology, 1965-66, University of Chicago), when we talk of the rise of urbanization, we must also talk of the process of peasantization. We must recognize there must be a process of change of agricultural towns from more or less self-sufficient units, to units that are part of the more complex symbiosis of civilization. At the time of the Conquest the Cuicatec *cacicazgos* were part of such a complex system, but they may not have always been.

In summary, then, at the time of contact with the Spaniards, the Cuicatec *cacicazgos* can be described as part of a nested set of systems with three hierarchical levels. The highest level is the Mesoamerican macro-system. The lowest level was that of the individual *cacicazgos*. These consisted of one or a few contiguous towns. The Cuicatec *cacicazgos* can be divided on the basis of their distribution in the environment into two groups, highland and lowland *cacicazgos*.

The highland towns were mostly based on *temporal*, or rainfall, agriculture. They had some small irrigation, and produced crops mostly for their own consumption. They produced some special highland products, such as wood, pulque, and hunting products. Some towns had special resources like salt and gold. Highland towns often had *sujeto* settlements that represented population that had hived off from the parent nucleus (Hunt 1972).

Lowland towns were invariably entirely dependent on irrigation for their agriculture. They could be described best as agricultural specialists. They grew surpluses of subsistence crops, such as corn, beans, squash, and chiles, which they traded to the highland communities. They also grew high-value specialty crops like fruit, which they traded through the region, as far off as to the Valley of Oaxaca and the Mixteca Alta.

The lowland towns lay in the Cañada, which was a major Mesoamerican north-south route. As a result, most of the markets were in the Cañada towns. Cuicatlán may have actually had a class of people who functioned as merchants, carrying goods up to trade with the highland towns.

The lowland towns may have been somewhat more stratified than the highland towns, although the evidence for this is weak. However, the lowland towns tended not to have *sujetos* outside their immediate area. Their size was limited to one or more contiguous irrigation system, the largest of which was probably the Cuicatlán-San Pedro system, in which San Pedro was a *barrio* of Cuicatlán. None of the *cacicazgos* seem to have been able to form a political unit that was very stable across even a small gap of unirrigated territory. Cuicatlán and Quiotepec, for example, were separated by only a few kilometers, but they never really formed a stable political unit.

Most wars waged by the lowland communities are characterized by Eva Hunt (1972) as "defensive," since these towns are tied down quite rigidly by the conditions that create their irrigation systems—the existence of small, perennial streams with which to irrigate, and flat land on which to deploy it. In contrast, the highland towns were not so limited on where to settle, and their "wars" may have been more in the nature of predatory raids.

The second level system is the Cuicatec ecosystem as I have defined it. There is no evidence that this level of integration was ever represented by a single *political* integration of any importance.

There is considerable economic interdependence between the *cacicazgos* in the region. This consists in part of "symbiotic" relations as described by Sanders (1956) and in part of specialization of the towns. In part, these economic integrations are paralleled and reinforced by the kin ties of the *caciques* that are dictated by class endogamy among the *caciques*.

However, the symbiotic, economic, kin, and political relationships crossed the boundaries of the ecosystem. Cuicatec towns were often attached politically to non-Cuicatec towns of neighboring ecosystems, and Cuicatec *cabeceras* had non-Cuicatec *sujetos* on the boundaries. This is probably the reason for the linguistic ambiguity of the borderline *cacicazgos*. Some of the border towns seemed to interact more with towns outside the ecosystem. Teutila, which was a Cuicatec cacicazgo with Mazatec and Chinantec *sujetos*, traded down the east side of the sierra, to the Veracruz lowlands, and to the west to non-Cuicatec towns north of the Cuicatec Cañada. However, most of the Cuicatec towns did tend to have closer relations with other Cuicatec towns within the ecosystem than with other towns outside the ecosystem.

Finally, at the time of the reporting of the sixteenth-century sources, the Cuicatec ecosystem was a part of the Aztec "empire." It formed part of the base for the expansionistic Aztec state, supporting the secondary centers of the empire that allowed the empire to reach beyond the immediate zone of influence of the Valley of Mexico. One can speculate on the effect of incorporation into this system on the structure of the *cacicazgos* in the Cuicatec ecosystem. We have some evidence that they had to organize part of their activity in response to the need to produce tribute beyond the subsistence needs of the community. At any rate, any analysis based on the historical sources of the Cuicatec *cacicazgos* must take into account the fact that they were not independent political units but rather were part of the Aztec system at the time they were recorded in history. These, then, were the Cuicatec *cacicazgos* at the time of the Spanish Conquest.

Chapter 3

The Spanish Conquest: The Cuicatec in Culture Contact

Because the Spanish encounter with the Mesoamerican civilizations was fairly recent, rich historical sources allow the reconstruction of the Mesoamerican civilizations that the Spaniards discovered. However, at the same time that the Spaniards recorded the civilizations, they were busy making important changes in it. The impact of the Spanish Conquest resulted in profound changes in many aspects of Mesoamerican civilization and in the ecosystems that were part of it. On occasion the same individuals to whom we owe the most for recording descriptions of Mesoamerica were those most active in destroying pre-Hispanic documents.

One regrets the loss of this kind of information. Some changes affected the civilization almost immediately, so that even the first reporters saw Mesoamerica when it had already begun to change. We can learn from the changes that resulted from the culture contact between the Spaniards and Mesoamerican Indians. Because the Conquest affected variables in the ecosystems selectively, it highlighted the workings of these variables within the ecosystem. As a result, we gain a better understanding of the Cuicatec ecosystem because of the changes brought about by the Spanish Conquest.

The Spanish Conquest is an interesting case of culture contact. There is no gainsaying the incredible achievements of the conquistadores in overthrowing vast political units. As a result of this Conquest the power at the top of the conquered political units changed dramatically. However, it is clear both from the history of the sixteenth-century Conquest and the culture that has survived to the present, that contact did not result in simple conquest and replacement of one culture with another. The two cultures—the Indian and the Spanish—that came into contact 400 years ago have been interacting to the present. In order to analyze the workings of the Cuicatec ecosystem over this period of time it is necessary to first understand this interaction between these two cultures.

Culture Contact and Schismogenesis

In order to analyze this culture contact, I have followed Gregory Bateson's discussion in his article "Culture Contact and Schismogenesis" (Bateson 1935; reprinted, 1972). He suggests that we consider first the group *before* culture contact.

> I suggest that we should consider under the head of "culture contact" not only those cases in which the contact occurs between two communities with different cultures and results in profound disturbance of the culture of one or both groups; but also cases of contact within a single community, e.g., between the sexes, between the old and young, between aristocracy and plebs, between clans, etc., groups which live together in approximate equilibrium. [Bateson 1972:64]

I would like to sketch very quickly, as he suggests, the situation of the Cuicatec ecosystem as a case of culture contact *before* the Spanish arrival.

Now, as Bateson points out, the relationships of two groups existing in dynamic equilibrium can be in two forms, symmetrical differentiation and complementary differentiation. While one or the other of these will characterize a relationship, a symmetrical relationship may have complementary aspects, and vice versa. This may contribute towards an equilibrium in the relationship.

> It is possible that, actually, no healthy equilibrated relationship between groups is either purely symmetrical or purely complementary, but that every such relationship contains elements of the other type. It is true that it is easy to classify relationships into one or the other category according to their predominant emphases, but it is possible that a very small admixture of complementary behavior in a symmetrical relationship, or a very small admixture of symmetrical behavior in a complementary relationship may go a long way toward stabilizing the position. [Bateson 1972:70]

There were both symmetrical and complementary aspects to the relationships within the pre-Hispanic Cuicatec ecosystem, and there were also complementary and symmetrical relationships in the relationship of the Cuicatec ecosystem to the macrosystem around it.

Within the individual Cuicatec towns which form the Cuicatec ecosystem, the relationship of the *cacique* and the nobles to the rest of the citizens is complementary. Bateson describes this relation:

> To this category we may refer all those cases in which the behavior and aspirations of the members of the two groups are fundamentally different. Thus members of group A treat each other with patterns L, M, N, and exhibit the patterns O, P, Q, in dealings with group B. In reply to O, P, Q, the members of group B exhibit the patterns U, V, W, but among themselves they adopt patterns R, S, T. Thus it comes about that O, P, Q is the reply to U, V, W, and vice versa. [Bateson 1972:68]

Thus all the nobles exhibit (roughly) the same behavior patterns toward each other and distinct patterns towards commoners. The commoners similarly behave towards each other with one set of behavior, but they act towards the nobles with a set of behavior patterns that is different both from how they behave to other commoners and from how the nobles behave to the commoners, or to each other.

Another kind of complementary relationship within the Cuicatec ecosystem is that of the exchange of highland products for lowland products. It does not matter to Bateson's argument whether these complementarities are dictated by environmental considerations or not. In the Mesoamerican case, some of them are. However, the interchange of food products from the Cañada and woven cotton products from the highlands in the Cuicatec case, were also part of this complementary pattern.

This kind of complementary pattern is stated by Bateson to be a contributing factor to equilibrium.

> It is certain that, as in the case quoted above in which group A sell sago to B while the latter sell fish to A, complementary patterns may sometimes have a real stabilizing effect by promoting a mutual dependence between the groups. [Bateson 1972:70-71]

Eva Hunt (1972) suggests that there was a symmetrical relationship between the nobility in the towns of the ecosystem in the form of village exogamy and class endogamy. A symmetrical relationship, according to Bateson, is:

> the individuals in two groups A and B have the same aspirations and the same behavior patterns, but are differentiated in the orientation of these patterns. Thus members of group A exhibit behavior patterns A, B, C in their dealings with each other, but adopt the patterns X, Y, Z in their dealings with members of group B. Similarly, group B adopt the patterns A, B, C among themselves, but exhibit X, Y, Z in dealing with group A. Thus a position is set up in which the behavior X, Y, Z is the standard reply to X, Y, Z. [Bateson 1972:68]

The relationships of the Cuicatec ecosystem to those around it, such as those of the Mixtec highlands or of the Valley of Tehuacán, were symmetrical or complementary in the same way as the relations within the ecosystem. The exchange of complementary products, such as fruit and salt, paralleled the symmetrical political relationships. The nobles' exogamous relations probably extended across the linguistic boundary by which we have defined our ecosystem.

The Cuicatec ecosystem was also part of an important complementary relationship that stood above these predominantly symmetrical relationships. This is its relationship with the Aztec "empire." This relationship is analogous to the complementarity between the nobles and the commoners in the Cuicatec ecosystem (or, as I have suggested, the noble/commoner relationship paralleled that between the Aztec and the Cuicatec systems). Two factors of the macrosystem affecting the Cuicatec ecosystem were that the Cuicatec had to send tribute and, at times, men to serve the Aztec system. At the same time, the Aztec made decisions affecting the running of the Cuicatec ecosystem, including assessments of tribute, and imposition of the Pax Azteca. Thus, the Cuicatec ecosystem represented a small complementary system which, in turn, was part of a larger complementary system under the Aztec. This larger system was expanding rapidly when the Spaniards came.

The Spanish Conquest of the Cuicatec Region

Jorge Fernando Iturribarría (1955) argues that there never was a conquest of Oaxaca, in the sense of a military campaign or campaigns in which, by superior force, technical ability, or tactics, the Spaniards forced the Indians in Oaxaca to give up. The Zapotec, in part to save themselves from a weak position in a war they had with the Mixtec, allied themselves more or less voluntarily. Iturribarría attributes this to their religious fatalism. He argues that they believed the omens that are supposed to have preceded the Spanish Conquest which showed the Spaniards to be the true successors of Quetzalcoatl. There were rebellions of towns and even larger regions in the state later, but in general the Iturribarría thesis holds true. The Mixtec, Zapotec, and Chinantec did ally themselves with the Spaniards in opposition to the Mexica, although they may have had political as well as religious motives.

The Cuicatec region is not very prominent in the early chronicles of the Conquest. Most evidence for its conquest is indirect. Iturribarría (1955) and Altolaguirre (1954) say that the first Spaniard sent to the Valley of Oaxaca was Francisco Orozco in 1521, but it is not clear what route he followed. Iturribarría points out that it took him fifty-five days to get to Etla from Tepeaca, which suggests that "hizó rodeos y extremó las precauciones en prevision de un asalto mixteca" (Iturribarría 1955:56). In these "rodeos" he may well have passed through the Cañada.

Herrera says that,

> el primero que entró á pacificar esta provincia [Oaxaca] fue Juan Nuñez de Mercado, Año de 1522 por comisión de D. Hernando Cortes: de alli se embio Gente de Guerra a servir al Tei Quatimoc en la defensa de Mexico, quando Don Hernando Cortes la sujeto. [1730, I:19-20]

Further on, in describing the "Provincia de los Çapotecas, i Cuioatecas, i otros" he says further,

> En el Pueblo de Tecomauaca, que está en el Camino Real de Guaxaca á Mexico, iendo Monteçuma á dar Batalla á los Indios de Zapotitlán, i pesandole, que se llevase en su exercito mas cuidado del regalo, i de lo que se avia de comer, que de las Armas, con que avian de pelear, mando quebrar todas las Xicaras, i Tecomaques, que son Vasijas de aqui quedó este nombre de Tecomauaca: i esta Tierra pacificó, por mandado de Hernando Cortés, Juan Nuñez de Mercado. [Herrera 1730:Dec. 3, Lib. 3, Cap 15:101, Año de 1522]

The very lack of historical notice of the conquest of the Cuicatecs indicates the quietness with which they passed from the Aztec to the Spanish yoke.

The Impact of the Spanish Conquest

Bateson (1972:64-65) tells us that

> If we consider the possible end of the drastic disturbances which follow contacts between profoundly different communities, we see that the changes must theoretically result in one or other of the following patterns:
> (a) the complete fusion of the originally different groups
> (b) the elimination of one or both groups
> (c) the persistence of both groups in dynamic equilibrium within one major community

Now, all through the post-Hispanic history of Mesoamerica, and demonstrably within the Cuicatec ecosystem at the present, there have been two major cultural patterns. The first pattern is called "Indian." "Indians" are the cultural and historical descendants of the pre-Hispanic groups. The other pattern was originally that of the Spanish colonists. Later those who followed this pattern were called variously Spanish, creole, mestizos, participators in the national culture, etc. I will refer to this pattern as "Spanish." Since this opposition has persisted, alternative (c) above, "the persistence of both groups in dynamic equilibrium within one major community" describes the result of this culture contact.

Initially, the Spaniards simply replaced the Aztec at the top of the complementary structure of Mesoamerica. However, as the first century of Spanish contact passed, there was an interesting change in this structure, at least in its interaction with the Cuicatec ecosystem.

First, there was a gradual disappearance through the century of the native noble class. Throughout the sixteenth century the upper classes were clearly being submerged into the mass of Indians. This was in part due to the immense impact of depopulation, which wiped out the Indian aristocracy as well as their constituency, and it was due in part to deliberate colonial policy. The Spanish rulers gradually decreased the tribute collected by the *caciques*, and phased out these *señores naturales*. Eva Hunt cites court cases on inheritance rights of *caciques* from a 1562 document from the Cuicatec region, but the *Libro de las Tasaciones* for the end of the same decade refers to *regidores*, not a *cacique* or *señor* of any kind, who held the keys of the *caja de comunidad* and disbursed the accumulated money. The *relaciones* of 1579 and 1580 of the Cuicatec towns mention several ways, especially in diet, that the distinctions of the nobles from the commoners had been eroded. So, as Chevalier puts it,

> en todo caso, hay un hecho muy claro: el prudente virrey Velasco reglamentó, limitó y aun suprimió gran cantidad de rentas pagadas por esos indios «terrazgueros», y sus sucesores lo imitaron . . . . Debido a estas razones y a otras más, la aristocrácia se encontraba muy de capa caída antes de finalizar el siglo XVI. [Chevalier 1956:166]

The complementary structure of the Indian towns was collapsed to a symmetrical one. At the same time, the Spaniards were establishing their own complementary structure over the towns. This first coexisted with the nobles and priests of the Indian hierarchy and then replaced it. Once the Spaniards had established their domination, the first thing they did was to divide up the territory and assign it to administrators. In the Cuicatec area this was originally done in the form of grants of encomienda.

The encomienda was originally intended as "a benign agency for Indian Hispanization" (Gibson 1964:58). The Indians granted to an encomendero owed him tribute and labor as they did to their *cacique*. However, he was in possession of the right only to these two things: the Indians' tribute and labor, not the land they inhabited. Theoretically, an encomienda was not inheritable. In fact the encomienda became the first means of exploitation of the Indians. From the first, tribute and labor that could be extracted by an encomendero were not regulated, and the Indian communities were in fact exploited quite unmercifully by the encomenderos.

This exploitation was quite heavy in the beginning of the sixteenth century. However, the encomiendas soon ran into heavy opposition, both from the church in its efforts to protect the Indians, and from the monarchy (Chevalier 1956; Gibson 1964). The effect of this opposition was a gradual limiting of the powers of the encomienda by regulation, which resulted in the attrition of the encomiendas.

The most common form of attrition was the escheatment of the encomienda to the royal crown on the death of the encomendero. Theoretically, the encomienda was not inheritable. The original encomenderos resisted this and managed to get some encomiendas extended to the second and even the third generation. However, most encomiendas rarely lasted this long, and all reverted after three generations. Encomiendas could also go to the crown for a number of reasons, including non-residence. The crown first limited the right to sell, then to inherit, and finally the encomenderos' right to unrestricted labor. By 1570 the encomendero class had essentially been defeated (Gibson 1964:63).

Five Cuicatec towns were in the hands of only one encomendero before they reverted to the crown. Four of these reverted at the death of the encomendero, and the fifth reverted because its encomendero left New Spain. Quiotepec probably belonged to Juan Nuñez Mercado (Gerhard 1973) and reverted to the crown in 1531 (*Libro de las Tasaciones* 1952:380). Teutila was assigned to Diego de Ordaz (Gerhard 1972) and at his death reverted to the crown in 1533 (*Libro de las Tasaciones* 1952:458). Tutepetongo and Tonaltepec were encomiendas of Juan Ochoa de Lexalde, "vecino de los Angeles" (*Libro de las Tasaciones* 1952:548-51; *Suma de Visitas* 1905; Spores and Saldaña 1973: No. 2115) from 1544 to 1556. In 1556, he died, but his wife and children still claimed the right to the tribute, in dispute with the Indians of these towns (*Libro de las Tasaciones* 1952:551). By 1580 they were a crown *corregimiento* under Melchor Suárez (a *vecino* of Antequera and one-time encomendero of Mixtepec) (Spores and Saldaña 1973: Nos. 972, 973). Tepeucila reverted to his majesty in 1544 "por ausencia de Luis de Cárdenas, que fue a la especeria" (*Libro de las Tasaciones* 1952:405).

Several of the other Cuicatec towns were passed from hand to hand very early (one gets the feeling the conquistadores were playing with huge tracts of land like monopoly money) but still reverted to the crown early without being successfully being passed on by any of the encomenderos to their families. Alpizagua (Dominguillo) was in encomienda for Juan Moreno (conquistador) and then reassigned to Gerónimo de Salinas in 1527 (Gerhard 1972). In 1544 it went to the crown (*Libro de las Tasaciones* 1952:381). Pápalo was the encomienda of Francisco de Ribadeo at first (Gerhard 1972), but was later in the hands of Francisco Casco and passed to the crown in 1541 when Casco died (*Libro de las Tasaciones* 1952:285).

Some of the towns were handed down from one or two generations, and a few even made it into the seventeenth century, but all ended up in the crown *corregimientos*. According to Gerhard (1972), Cuicatlán was originally assigned by Cortés to Juan Tirado, then by 1524 to Juan de Jaso, then reassigned by the first *audiencia* "perhaps to Tirado, who held it in the 1530s, followed by a son" (Gerhard 1972:306). By 1548, it was listed in the *Suma de Visitas* as being "en su majestad."

Atlatlauca was originally divided in half between two encomenderos (Gerhard 1972). One-half went to the crown in 1532 (*Libro de las Tasaciones* 1952:85) by order of the second *audiencia*. The other half belonged to Juan de Mancilla, who sold it to Juan Gallego in 1538 (Gerhard 1972). Juan Gallego had handed it on to his son by 1565 (Spores and Saldaña 1973: No. 156), who held it until 1597 (Gerhard 1972).

Nanalcatepec (Nacaltepec) was in encomienda under Melchor de San Miguel, who was also encomendero of Tequixtepec in 1560 (Gerhard 1972; Spores and Saldaña 1973: No. 972). According to Gerhard, soon after 1560 his son inherited. Either he or his son had their power limited in 1561 (Spores and Saldaña 1973: No. 973):

> 973. 1561 Nacaltepec. Encomienda. Ordenando que los naturales de este pueblo, que tiene en encomienda Melchor de San Miguel, no sean obligados a dar más aportación de su encomendero, si no del tiempo que corriere desde la data. Vol. 5. Fa. 197 vta.

Later it went to his (the son's) widow, María de Godoy, who also seems to have inherited Tequixtepec, since she is listed as encomendero in 1560 (Spores and Saldaña 1973: No. 1845), where she was accused of maltreating the natives. She is also listed as encomendero of another town (Spores and Saldaña 1973: No. 214). She held Nanalcatepec until her death in 1587 (Gerhard 1972) when it finally went to the crown.

Finally, Nacaltepec's neighbor, Cuitlahuistla (Cotahuixtla) was passed on apparently for at least three generations. It was originally granted to Gonzalo de Robles, then passed on to his son, García de Robles, who held it through 1580, and finally was held by Juan de Robles from 1597 to 1604 (Gerhard 1972). Gonzalo de Robles was also encomendero of Apoala and Xoquitipaque in 1551 (Spores and Saldaña 1973: No. 135).

Thus 7 ½ of the Cuicatec towns were crown *corregimientos* before the time of the *Suma de Visitas* (1548). These were Tepeucila, Teutila, Quiotepec, Alpizagua, Pápalo, Cuicatlán, Camotlán, and half of Atlatlauca. Of the remaining towns, Tutepetongo and Tonaltepec passed to the crown before the *relaciones* of 1579-80. The half of Atlatlauca and the encomienda of Nacaltepec did not survive the sixteenth century. Only that of Cuetlahuixtla (Cotahuixtla) lasted into the seventeenth century.

As has been pointed out, Tutepetongo and Tonaltepec, although Cuicatec, were part of the Mixteca politically. In addition, they were involved in the production of grana (cochineal), which may be a reason why their encomenderos held onto them longer. The *Libro de las Tasaciones* lists changes in their assessment so that they supply grana instead of other assessments. The other three towns, Nacaltepec, Cotahuixtla, and Atlatlauca, lie on a line along the Camino Real from the Etla arm of the Valley of Oaxaca to the Cañada and are all in the same region. First, this is the end of the Cuicatec area that is accessible to an area populated by a sizable population of Spaniards in the Valley of Oaxaca. Secondly, at least one of the people involved, the second Juan Gallego, was considerably involved with the Cañada, particularly with the southern end of it. He was the encomendero of half of Atlatlauca. At the same time he was the *corregidor* of Cuicatlán and is the one who wrote the *relación* in 1580. It was he, or his father, who in 1560 made the *relación* describing the condition of the Camino Real and recommended that it be straightened (Spores and Saldaña 1973: No. 1108) and he also had a sugar *ingenio* (refinery) at San Juan, in 1590, near Huajolotitlán (Huitzo) on the route to Atlatlauca (Spores and Saldaña 1973: No. 1304). Sugar was probably one of the crops he extracted from his half of Atlatlauca.

One would expect that encomiendas, in spite of the considerable pressure described in Gibson (1964) and Chevalier (1956) would have survived longer in this backwater than in the Valley of Mexico, for instance. That they did not in general is probably a function of the fact that the Cuicatec area was not (relatively) a very rich one for the first exploitation of the early Spaniards. First, the Cañada was too dry for successful large-scale cattle exploitation, and the highland towns were almost all described as being too cold and of too broken terrain even for *ganado menor*. Since this was the first thrust of the big haciendas, this area would have gone relatively unscathed. Secondly, since initially there were relatively few Spaniards, concentrated in a few centers, they simply would not have reached this relatively out-of-the-way part of Mesoamerica.

This point is important when we consider the *corregimientos* that succeeded the encomiendas. As Gibson (1964) points out, the *corregimientos* were the true successors of the encomiendas, in that they were the instruments of exploitation in many ways. However, from the descriptions one gets the impression that no one was very interested in the Cuicatec region. Few Spaniards ever did more than pass through the area in transit on the Camino Real. Padre Ponce only mentions Spaniards in 1568 in Cuicatlán: "moran en aquel pueblo algunos españoles, y allí está de asiento el clérigo" (1967:19). The rest of the towns are described as poor, and one gets the impression that it was all that the Indians in the other towns could do to even scrape up the food to feed the Padre. Gerhard says of Atlatlauca that,

This unimportant post, suffragan to the alcaldía mayor of Antequera in 1552-1603, had little to attract office seekers. The viceroy in 1591 ordered the C[orregidor] not to live there because the Indians were too few and too poor to support a magistrate; justice was to be administered by the A[lcalde] M[ayor] of Antequera. [Gerhard 1972:54-55]

The history of Catholic priests parallels that of the encomenderos. Following the establishment of a fairly big religious establishment, the Spanish religious personnel residing in and working in the Cuicatec region seems to have been reduced quickly.

The earliest reference to priests (*religiosos*) in the Cuicatec towns is an order that housing be made for friars in Teutila in 1543 (Spores and Saldaña 1973: No. 1715). Gerhard says that the Dominican monastery at Teutila was established in 1560. And Spores and Saldaña have an order for the sustenance of these priests in 1563 (1973: No. 1867). Gerhard says that the monastery was later secularized. By 1766 Ajofrín (1959) says that the monastery was in ruins when he passed through, though he says the monastery was Franciscan.

By the 1560s we have statements that all of the towns had some kind of priest ministering to them. Cuicatlán was said to be the clerical seat in 1568 (Ponce 1967). The *relación* of 1580 says that there was a priest in Cuicatlán who spoke Cuicatec and administered to the towns. Tutepetongo had a *vicario* at the time of its *relación*, while Tonaltepec was administered by a priest from Zozola.

Two of the towns had priests who were responsible for several towns. Tepeucila seemed to have always been administered as part of the *doctrina* of Pápalo. The priest from Pápalo visited Tutepetongo and Cuicatlán (*Relación de los Obispados del Siglo XVI* 1904). Tepeucila and Pápalo were assessed for the support of priests in 1564 (Spores and Saldaña 1973: No. 1787). The priest in Atlatlauca was also responsible for Cuetlahuixtla (Cotahuixtla) and Nacaltepec. According to Gerhard (1972), Alpizagua (Dominguillo) was also administered from Atlatlauca.

So the Cuicatec area received a great deal of attention from the priests. The Indian religion was almost immediately suppressed, and the Indians converted to Catholicism. However, by the end of the sixteenth century there were only a few priests who were responsible for the widely-scattered Cuicatec towns.

The priests and encomenderos first paralleled, then replaced the Indian nobles. However, by the end of the sixteenth century we find the Indian communities deprived of their noble class, but substantially neglected by the Spanish authorities who replaced them. As a result the Indian villages changed from a complementary structure to a symmetrical one. The Spaniards then tended to concentrate their political and religious authority in a few individuals who traveled over the entire region.

This meant that the Spaniards had taken over all the complementary, hierarchical relations. The Indians were reduced to symmetrical structures within their towns. Any complementary structure in which the Indians participated involved the Spaniards with the upper hand in the relationship. The Indian towns were part of a larger complementary structure with the Spaniards. The Indian had become a kind of peasant.

To summarize, then, we find that the result of the contact between these two cultures was the formation of an opposition between the two groups and the establishment of two patterns of existence, the "Indian" and the "Spanish" patterns. Both have survived, though with many changes, to the present, but they are structurally different. The Indian culture pattern is symmetrical, or egalitarian, confined to and oriented to the local area, and oriented to subsistence rather than profit.

The Spanish communities, as we have seen, abrogated to themselves all hierarchical relations and power. They first controlled the national or Mesoamerican-wide system (replacing the Aztec) but also even on the local level kept all power in Spanish hands. The local Spanish communities, unlike the post-Hispanic Indian pattern, but like the pre-Hispanic Indian pattern, were microcosms of the national system, with complementary relationships within the structure of each community. Spanish communities were oriented towards the macrosystem, on national and international relations, while the Indian communities were oriented inwards.

The fact that the Indian pattern survived for 400 years as one side of this opposition shows that there must have been some forces that held the forces for schismogenesis in equilibrium. These must have included some sort of leveling force to keep classes from arising. There must have been other forces that prevented the communities from simply changing to match the Spanish pattern, a logical possibility of the schismogenesis.

The forces for equilibrium and maintenance of the symmetrical structure within the Indian communities have been described by Wolf (1957) in his description of what he calls "closed corporate peasant communities." The second negative feedback lay in the nature of the complementary structure of the relationship of the Indian communities to the Spanish macrosystem. While at the local level the forces for schismogenesis would seem to be inevitable, when the Indian communities are seen in the larger equilibrium one sees how stable they are. One must see the post-Hispanic Indian communities as part of a larger system that defines them as peasants.

The structure of this opposition was the structure of the human part of the Cuicatec system. The interaction of the natural and the cultural subsystems in the ecosystem dictated the fate of the new elements introduced by the Spaniards.

Just as the arrival of the Spaniards resulted in a reorganization in the cultural system, so too did it have an important impact on the natural system. This was the effect of the Spanish arrival on the Indian population. This effect, like the structure we have been discussing, has affected the Cuicatec ecosystem for its entire post-Hispanic history.

Post-Hispanic Population Change and the Cuicatec Ecosystem

The most important and long-lasting effect of the Spanish Conquest was the massive depopulation that afflicted Mesoamerica after the Conquest. This process of depopulation and later recovery is still going on. The process has had important effects on the working of the Cuicatec ecosystem as the population values changed over time.

Many explanations were offered for the fall in population. Modern explanations have ranged from wars of conquest, the death of Indians from their hard labor under the Spanish encomenderos, and culture shock. To these Gerhard adds those attributed by witnesses contemporary to the depopulation, of "Mistreatment or too benign treatment, starvation or rich foods, despondency and drunkenness, unaccustomed liberty, the wearing of clothes, the use of beds, too frequent bathing, divine retribution, etc." (Gerhard 1972:23). As he points out, the chief villain was the introduction of epidemic diseases such as smallpox, measles, and malaria, to which the Indians had no natural resistance. Estimates of the extent of the impact of this depopulation range from half the population being wiped out to a reduction to one-twentieth of the preceding population.

Estimating Population for the Cuicatec Ecosystem

From the late nineteenth century on, we have fairly good census data from the Cuicatec area. These estimates can usually be taken at face value. There are a few cases where population figures for one date for one or another place are obviously incorrect. Usually this is due to inaccurate reporting, or, occasionally, obvious typographical errors. In the censuses population is either given in males, females, and the total number of people, or with further breakdowns into categories of age or cultural category (e.g., Indians vs. non-Indians).

In the earlier sources, calculation of population is more difficult. Total population is usually not given, in part because the government agencies and the other commentators recording the population were not interested in the total population. Instead, much of the population data either comes from tributary records, or other reports related to the calculation of tribute, or from ecclesiastical sources, which often reflected the categories used by the treasury tribute-collecting records.

Consequently the sources from which population can be estimated in the sixteenth century use and report population in many different units, which must be multiplied by some factor to approximate the total population. The units which occur for the Cuicatec sources are: *personas* (literally, "persons," but see Cook and Borah 1960), *casas* ("houses"), *vecinos* (literally "neighbors"), *casados* ("married men"), *muchachos de quince años de abajo* ("children fifteen years or younger") and *tributarios* ("tributaries"). Later sources often refer to *familias* ("families"). Finally, if one wants, one could try to derive approximate population estimates from the amount of tribute paid in a town, on the assumption of a unit tribute payment per capita. Clearly all of these involve the calculation of ratios of vecinos, casados, tributarios, and even pesos of tribute to total population.

Fortunately for the novice, prioneering studies by Cook and Simpson (1948), Cook (1949, 1968), Borah and Cook (1960, 1963), and Cook and Borah (1960) have been made. These studies attempt to deal with these problems of deriving population from the early sources for Mesoamerica. The studies do not give a single answer to exactly which factors to use with each term. Rather they represent an evolution, over time, of attempts to analyze the historical estimates of population. The Cook,

Simpson, and Borah estimates are based on comparisons of statements where more than one term is used.

After reading and being somewhat confused by the various elaborate calculations for deriving conversion factors for the sixteenth century and later, I prepared a table of the Cuicatec data (Table 2). I listed all the towns from the Cuicatec area, and the kind of population information available for each, along with the date and source for each available statement.

In the data that precede 1560, the estimates showed an even decline over time of the raw statements, whether amount of tribute, or the number of casados, tributarios, etc. The two most important sources for this segment of time, the *Libro de las Tasaciones* and the *Suma de Visitas*, clearly show this. The *Libro de las Tasaciones* begins with tribute assessments in material goods in the 1530s, then changes these to silver or gold (currency) and then records a series of diminutions of tribute, which presumably represents a trend in diminution of the population, since it is hard to think that the Spaniards allowed the tribute to be lowered out of altruistic concern for the Indians.

However, in the late 1560s the tribute recorded in the *Libro de las Tasaciones*, and the number of tributaries recorded, in the *Libro de las Tasaciones* and in other sources, abruptly jumps high above the preceding figures. This leads to a problem of interpretation. If we are to use a single figure for conversion of tributaries to population, then we must postulate a sudden jump in population in the decade of the 1560s, as well as an abrupt increase in the ability to pay tribute, or at least an increase in the assessment. Or, if we are to postulate a change in the ratio of tributaries to the total population (that is, that more people were counted as tributaries than previously) then we must derive two factors, one for before the change in policy and one for after.

However, the relationship between these two factors is, in part, the plaything of our theories about the population trends. Thus, a lot of figures and figuring (Cook and Borah 1960; Borah and Cook 1960, 1963; Cook 1968) involve calculations to make the figures generate a smooth curve of some kind.

The reason for this change can be found, both from internal and external evidence. It turns out there was indeed a change of policy in the late 1560s. Gibson (1964) explains that a change was going on in this period, as a result of pressure from Spain for more income. This was in particular the result of the Valderrama *visita*.

> The change received a special impetus following the Valderrama visita, the express purpose of which was to increase the tribute yield. Whereas Diego Ramírez in the 1550's and other royal agents had tended to abide by the original Indian criteria of exemption, Valderrama uniformly rejected these criteria. In the far-reaching changes of the 1560's the sub-macegual classes were almost without exception classified as tributaries, and even caciques were sometimes placed on the tribute rolls. [Gibson 1964:200]

Borah and Cook (1960) discuss these changes in their analysis of the *Suma de Visitas*. They point out that before the reclassification, substantial portions of the Indian population were exempt from consideration as tributaries. Tributaries were originally calculated as married free males. Not all males fit this category. Slaves, and other categories lower than the maceguales were exempt from tribute, as were merchants, and some nobles. In contrast, after the 1560s almost everyone living outside his original family except the *cacique* was liable for tribute. Married men counted as full tributaries, but widows, widowers, and single men who owned their own land or had their own households counted as half-tributaries.

As a result the ratio of tributaries to total population changed substantially. Having recognized that this is a problem, we then return to the various studies on the population of Mexico (Cook and Borah 1960, 1963; Borah and Cook 1968; Cook 1968), and apply the various factors that they have used in these works, to see which ones seem to make sense with the Cuicatec data. The factors by which to multiply tributaries in order to derive total population range from 3.2 to 5.0. The criteria I have ended up using are those delineated in Borah and Cook (1960), Cook and Borah (1960), and Cook (1968).

In these I have accepted the following rules for making transformations from the statements in the sources to the total population. First, for statements of "personas" I have added ⅑ of the figure to the total, for the people under the age of three who would not have been counted yet as personas. This figure of ⅑ is also added where the population is broken down into "casados" and "muchachos de 15 años de abajo," as in the *Suma de Visitas*. I have accepted the Cook and Borah (1960) argument that casados, tributarios, and vecinos are generally

TABLE 2
CUICATEC POPULATION DATA

Date	Source	Statement of Population		Estimate
		Alpizagua (Dominguillo)		
1544	Libro de Tasaciones	32 "pesos de oro Común" tribute		
1548	Suma de Visitas	27 personas		
		12 cassas		
		10 cassas de advenedizas		
		10 tributarios	(×3.3)	63
		"poblezuelo"		
1568	Padre Ponce			
1569	Relación de los Obispados	30 indios tributarios	(×2.8)	84
1745	Theatro Americano	36 familias	(×6)	
			Men	Women
1882	Censo		188	163	351
1883	Cuadros Sinópticos		192	166	358
1900	Censo				340
1910	Censo				418
1921	Censo				402
1930	Censo				376
1950	Censo				392
1960	Censo				385
		Atlatlauca		
pre-Spanish	Relación (1580)	"de veynte partes no ay la una" 20 × 700 indios cassados = 14,000 indios cassados		46,200 (too high)
1532	Libro de las Tasaciones	40 "tejuelos de oro de 3 pesos 1 tomin cada uno" for ½ of Atlatlauca (= 125 pesos)		
1541	Libro de las Tasaciones	½ of Atlatlauca paid 366 pesos per year		
1544	Libro de las Tasaciones	120 pesos de oro		
1548	Suma de Visitas	(with sujetos) 658 tributarios (×3.3)		2,171
1550	Libro de las Tasaciones	93 pesos, 3 tomines, 6 granos tribute		
1560	Tributos de Pueblos de Indios	(with Marinalco) ½ its tribute is "1000 pesos y toldillos"		
1564	Libro de Tasaciones	(with subjects) 1131 pesos, 9 granos de oro común, 422.5 fanegas of corn 845 tributarios (with some sujetos) (×2.8)		2,366
		107 ½ tributarios (in other sujetos) (×2.8)		301
1569	Relación de los Obispados	800 indios tributarios (×2.8)		2,240
1580	Relación de Atlatlanca	(with Malinaltepec) 700 indios	(×3.3)	2,140
1745	Theatro Americano	68 familias	(×6)	408
			Men	Women
1930	Censo	(municipio)	481	517	998
		(cabecera)	202	196	398
1960	Censo	(municipio)	405	444	849
		(cabecera)	194	211	405
		Camotlán		
1548	Suma de Visitas	40 casas		175
1548	Suma de Visitas	53 tributarios	(×3.3)	
		69 mugeres		
		63 muchachos		
		185	(×1%)	205

TABLE 2 CONTINUED
CUICATEC POPULATION DATA

Date	Source	Statement of Population			Estimate
		Chapulalpa			
1882	Censo		Men 143	Women 135	278
1883	*Cuadros Sinópticos*		144	138	282
1900	Censo				626
1910	Censo				426
1921	Censo				599
1930					898
1940	Censo				678
1950	Censo				546
		El Chilar			
1766	Ajofrín	"otro pueblito mas pequeno" (than San Pedro Chicozapotes)			
1882	Censo		Men 107	Women 91	198
1883	*Cuadros Sinópticos*				196
1900	Censo				350
1910	Censo				277
1921	Censo				460
1930	Censo				434
1940	Censo				644
1950	Censo				621
1960	Censo		423	411	834
		Chiquihuitlán			
1882	Censo		Men 921	Women 1,206	1,947
1883	*Cuadros Sinópticos*		928	1,037	1,965
1900	Censo				2,343
1921	Censo				2,711
1930	Censo				2,951
1940	Censo				3,075
1950	Censo				3,593
1960	Censo	(cabecera)	1,368	1,435	2,803
		(municipio)	1,537	1,594	3,131
		Cotahuixtla			
1745	*Theatro Americano*	28 familias		(×6)	168
1882	Censo		Men 82	Women 87	169
1883	*Cuadros Sinópticos*		Men 84	Women 87	171
1900	Censo				261
1910	Censo				341
1921	Censo				368
1930	Censo				463
1940	Censo				492
1950	Censo				578

THE CUICATEC IN CULTURE CONTACT

Coyula

Year	Source	Description			
1548	Suma de Visitas	17 casas	(× 2 = 70)	(× 3.3)	116
		35 hombres casados	+20		
		20 muchachos de quinze años de abajo	90		
		30 familias		(× 1⅙)	100
1745	Theatro Americano			(× 6)	180
			Men	Women	
1882	Censo		141	134	275
1883	Cuadros Sinópticos		144	138	282
1900	Censo				369
1910	Censo				355
1921	Censo				381
1930	Censo				429
1940	Censo				397
1950	Censo				470
1960	Censo		279	257	536

Cuicatlán

Year	Source	Description			
1548	Suma de Visitas	140 casas	(× 2 = 644)	(× 3.3)	1,063
		322 casados	105		
		105 solteros	190		
		190 niños	939	(× 1⅙)	1,403
1560	Tributos de Pueblos de Indios	280 pesos			
1569	Relación de los Obispados	300 Indios Tributarios		(× 2.8)	840
1745	Theatro Americano	125 familias de indios			750
1766	Ajofrín	"algunas familias de Españoles" (besides Indians)			
ca. 1800	Thiéry de Menonville	200 familias		(× 6)	1,200
			Men	Women	
1882	Censo		640	615	1,255
1883	Cuadros Sinópticos		641	617	1,258
1898	Censo (distrito)	Spanish speakers	2,784	2,813	5,607
		Cuicatec sp.	3,888	4,151	8,039
		Chinantec sp.	938	1,120	2,058
		Total	7,620	8,808	15,704
1900	Censo				2,086
1910	Censo				2,193
			Men	Women	
1913	Esteva Cayetano	(municipio)	12,943	13,551	26,494
1921	Censo	(municipio)			2,203
1930	Censo	(cabecera)	2,180	2,171	4,351
			879	939	1,818
1940					1,894
1950	Censo				1,978
1960	Censo	(municipio)	3,725	3,600	7,325
		(cabecera)	1,239	1,216	2,456

Cuyaltepec, San Pedro

Year	Source				
1882	Censo		107	108	215
1883	Cuadros Sinópticos		110	110	220
1900	Censo				206
1910	Censo				292
1921	Censo				336

TABLE 2 CONTINUED
CUICATEC POPULATION DATA

Date	Source	Statement of Population			Estimate
1930	Censo				431
1940	Censo				467
1950	Censo				490
		Cuyamecalco			
			Men	Women	
1882	Censo		749	745	1,494
1883	Cuadros Sinópticos		758	756	1,514
1900	Censo				2,052
1921	Censo				2,142
1930	Censo	(cabecera)	1,165	1,192	2,357
	Censo	(municipio)	1,472	1,478	2,950
1940	Censo				2,016
1950	Censo				2,011
1960	Censo	(municipio)	2,084	2,168	4,252
	Censo	(cabecera)	1,177	1,211	2,388
		Cuytlaquiztlán			
1569	Relación de los Obispados	300 indios tributarios		(×2.8)	840
		Guaxospan			
1745	Theatro Americano	150 familias de indios		(×6)	900
		Guendalain (Valerio Trujano)			
			Men	Women	
1882	Censo		246	254	500
1883	Cuadros Sinópticos		244	256	500
1900	Censo		Men	Women	527
1910	Censo				544
1921	Censo				493
1930	Censo	(municipio)	316	302	618
	Censo	(cabecera)			424
		(name change to V. T.)			
1940	Censo				457
1950	Censo				630
1960	Censo	(municipio)	585	631	1,216
	Censo	(cabecera)	364	362	726
		Ixtepex			
1548	Suma de Visitas	451 yndios tributarios			1,488
1560	Tributos de Pueblos de Indios	250 pesos			
		Ixcatlán			
1745	Theatro Americano	500 familias de Indios		(×6)	3,000
		Nacaltepec			
1569	Relación de los Obispados	200 Indios Tributarios (Manalcatepeque)		(×2.8)	560
1745	Theatro Americano	33 familias de indios		(×6)	198

THE CUICATEC IN CULTURE CONTACT

Guadalupe Obos

Year	Source	Men	Women	Total
1882	Censo	141	147	288
1883	Cuadros Sinópticos	146	149	295
1900	Censo			500
1910	Censo			601
1921	Censo			633
1930	Censo	454	431	885
1940	Censo			800
1950	Censo			972

Oxitlán

Year	Source	Men	Women	Total
1882	Censo	54	59	113
1883	Cuadros Sinópticos	53	61	114
1900	Censo			56
1910	Censo			76
1921	Censo			128
1930	Censo			0
1940	Censo			71
1950	Censo			118
1960	Censo	65	69	134
1548	Suma de Visitas	44 cassas		
		128 personas casadas	(×3.3)	422
1560	Tributos de Pueblos de Indios	150 pesos		

Concepción Pápalo (Papaloticpac)

Year	Source	Details		Total
pre-1541	Libro de las Tasaciones	Tribute in produce, etc.		
1541	Libro de las Tasaciones	520 pesos gold		
1547	Libro de las Tasaciones	360 pesos gold		
1548	Suma de Visitas	319 cassas		
		507 indios tributarios	(×2 = 1,014)	1,673
		495 muchachos	−495	
			1,509	
			(×1%)	1,654
1560	Tributos de Pueblos de Indios	100 pesos		
1567	Libro de las Tasaciones	950 pesos gold, 380 fanegas de maíz		
		760 tributaries	(×2.8)	2,128
1569	Relación de los Obispados	600 indios tributarios	(×2.8)	1,680
1745	Theatro Americano	142 familias	(×6)	852

Year	Source	Men	Women	Total
1882	Censo	603	615	1,218
1883	Cuadros Sinópticos	604	622	1,226
1900	Censo			804
1910	Censo			1,363
1921	Censo			669
1930	Censo	443	457	900 (cabecera)
		1,034	992	2,026 (municipio)
1940	Censo			833
1950	Censo			928
1960	Censo	1,226	1,140	2,366 (municipio)
		415	405	820 (cabecera)

TABLE 2 CONTINUED
CUICATEC POPULATION DATA

Date	Source	Statement of Population	Men	Women	Estimate
		Reyes Pápalo			
1745	*Theatro Americano*	165 familias de Indios (with San Lorenzo) (×6)			990
1882	Censo		439	470	909
1883	*Cuadros Sinópticos*		444	469	913
1900	Censo				896
1910	Censo				1,048
1921	Censo				996
1930	Censo	(municipio)	550	572	1,122
1940	Censo				1,282
1950	Censo				1,558
1960	Censo	(municipio)	826	907	1,733
		(cabecera)	746	836	1,582
		San Andrés Pápalo			
1745	*Theatro Americano*	20 familias (barrio de Tepeucila) (×6)			120
1882	Censo		161	180	341
1883	*Cuadros Sinópticos*		161	182	343
1900	Censo				332
1910	Censo				311
1921	Censo				247
1930	Censo				222
1940	Censo				244
1950	Censo				283
		San Lorenzo Pápalo			
1745	*Theatro Americano*	165 familias (with Santos Reyes Pápalo) (×6)			990
1882	Censo		175	186	361
1883	*Cuadros Sinópticos*		176	189	365
1900	Censo				444
1910	Censo				413
1921	Censo				507
1930	Censo				498
1940	Censo				496
1950	Censo				560
		Santa María Pápalo			
1745	*Theatro Americano*	23 familias	(×6)		78
			Men	Women	
1882	Censo		368	402	770
1883	*Cuadros Sinópticos*		370	410	780
1900	Censo				853
1910	Censo				915
1921	Censo				828
1930	Censo				1,026
1940	Censo				1,020

THE CUICATEC IN CULTURE CONTACT

Year	Source	Description	Men	Women	Total
					714
1950	Censo		906	842	1,748
1960	Censo		488	452	940

Quiotepec

Year	Source	Description			
1548	*Suma de Visitas*	(municipio) (cabecera) 40 cases			
		91 hombres cassados	(×2 = 182)	(×3.3)	300
		50 muchachos de quinze años de abajo	50		
			232		
pre-1552	*Libro de las Tasaciones*	120 pesos, 80 camisas, 80 naguas per year			
1552	*Libro de las Tasaciones*	120 pesos de oro comun			
1568	Padre Ponce	"poblecito", vecinos "pobres y ... desapercibidos."			
1745	*Theatro Americano*	42 familias		(×6)	252
			Men	Women	
1882	Censo		126	152	278
1883	*Cuadros Sinópticos*		122	147	269
1900	Censo				305
1910	Censo				259
1921	Censo				358
1930	Censo				337
1940	Censo				364
1950	Censo				400
1960	Censo		229	210	439

San Pedro Chicozapotes

Year	Source	Description	Men	Women	Total
1766	Ajofrín	"Pueblito"			
1882	Censo		109	88	197
1883	*Cuadros Sinópticos*		109	89	198
1900	Censo				266
1910	Censo				394
1921	Censo				374
1930	Censo				363
1940	Censo				386
1950	Censo				523
1960	Censo		263	254	517

Santo Domingo del Río

Year	Source	Description	Men	Women	Total
1882	Censo		92	108	200
1883	*Cuadros Sinópticos*		94	110	204
1900	Censo				194
1910	Censo				240
1921	Censo				105
1930	Censo				66
1940	Censo				323
1950	Censo				253
1960	Censo	(part of the municipio of S. Pedro Teutila)	118	115	233

TABLE 2 CONTINUED
CUICATEC POPULATION DATA

Date	Source	Statement of Population	Men	Women	Estimate
		Tecomastlahuaca			
1882	Censo		71	73	144
1883	*Cuadros Sinópticos*		74	69	143
1900	Censo				216
1910	Censo				153
1921	Censo				113
1930	Censo				77
1940	Censo				20
		Teotilalpam			
1882	Censo		Men	Women	
1882	Censo		613	630	1,243
1883	*Cuadros Sinópticos*		617	633	1,250
1900	Censo				1,033
1910	Censo				1,125
1921	Censo				929
1930	Censo	(cabecera)	480	570	1,050
1940	Censo				843
1950	Censo				978
		Tepeucila			
pre-Spanish	*Relación de Tepeucila* (1579)	40,000 Indios Tributarios			112,000 (too high)
1544	*Libro de las Tasaciones*	384 pesos de oro en polvo			
1547	*Libro de las Tasaciones*	80 pesos de oro en reales			
1548	*Suma de Visitas*	142 tributarios		(×3.3)	469
		105 hijos			
					389
		(×2 = 284)			
		+105			
				(×1%)	432
1549	*Relación* (1579)	2,500 hombres tributarios		(×2.8)	7,000 (too high)
1560	*Libro de las Tasaciones*	242 pesos, 121 fanegas de maíz		(×3.3)	799
		242 Tributarios			
1560	*Tributos de Pueblos de Indios*	160 pesos			
1561	*Libro de las Tasaciones*	Maize changed to silver, = 287 pesos, 3 reales			
1566	*Libro de las Tasaciones*	360 pesos			
		288 tributarios		(×2.8)	806
1569	*Relación de los Obispados*	150 indios tributarios		(×2.8)	420
1579	*Relación* (1579)	100 hombres tributarios		(×2.8)	280
1745	*Theatro Americano*	106 familias de indios		(×6)	636
			Men	Women	
1882	Censo		255	280	535
1883	*Cuadros Sinópticos*		253	283	536
1900	Censo				588
1910	Censo				611
1921	Censo				585

THE CUICATEC IN CULTURE CONTACT

Year	Source	Description	Men	Women	Total
1930	Censo				700
1940	Censo				817
1950	Censo				868

Teponastla

Year	Source	Description	Men	Women	Total
1745	Theatro Americano	56 familias de indios		(×6)	336
1882	Censo		168	167	335
1883	Cuadros Sinópticos		171	168	338

Year	Source	Description	Men	Women	Total
1745	Theatro Americano	56 familias de indios		(×6)	336
1882	Censo		168	167	335
1883	Cuadros Sinópticos		171	168	338
1900	Censo				390
1910	Censo				213
1921	Censo				379
1930	Censo				412
1940	Censo				473
1950	Censo				564

Tequecistepeque

Year	Source	Description	Calculation	Multiplier	Total
1548	Suma de Visitas	15 casas			
		40 hombres	(×2 = 80)	(×3.3)	132
		20 muchachos	+20 = 100	(×1⅓)	111

Teutila

Year	Source	Description	Calculation	Multiplier	Total
1533	Libro de las Tasaciones	600 pesos de oro en polvo, 20 cargas de cacao			5,003
1548	Suma de Visitas	1516 tributarios	(×2 = 3032)	(×3.3)	
		1352 muchachos de 8 años arriba	1352 → 4384	(×1⅓)	4,871
1560	Tributos de Pueblos de Indios	1247 pesos, 3 tomines, 9 granos de oro común			
1566	Libro de las Tasaciones	525 fanegas, 3 almudes de maíz			
		1050.5 tributarios		(×2.8)	2,941
1569	Relación de los Obispados	120 indios tributarios		(×2.8)	336
1766	Ajofrín	300 familias de indios (town proper)		(×6)	1,800
		80,000 almas (jurisdicción de Teutila)			80,000

Year	Source	Locality	Men	Women	Total
1882	Censo	(Santa Cruz T.)	97	106	203
		(San Pedro T.)	187	214	401
1883	Cuadros Sinópticos	(Santa Cruz T.)	98	101	199
		(San Pedro T.)	184	212	396
1900	Censo	(Santa Cruz T.)			150
		(San Pedro T.)			445
1910	Censo	(Santa Cruz T.)			181
		(San Pedro T.)			467
1921	Censo	(Santa Cruz T.)			169
		(San Pedro T.)			581
1930	Censo	(Santa Cruz T.)			212
		(San Pedro T.)			773
1940	Censo	(Santa Cruz T.)			239
		(San Pedro T.)			862
1950	Censo	(Santa Cruz T.)			275
		(San Pedro T.)			1,162

TABLE 2 CONTINUED
CUICATEC POPULATION DATA

Date	Source	Statement of Population			Estimate
1960	Censo	(municipio)	1,404	1,454	2,858
		(S. Pedro cab.)	630	707	1,337
		(Cerros de Olla)	16	24	40
		(San Juan T.)	56	46	102
		Tlacolula			
			Men	Women	
1882	Censo		203	236	439
1883	*Cuadros Sinópticos*		206	237	443
1900	Censo				445
1910	Censo				507
1921	Censo				476
1930	Censo				564
1940	Censo				678
1950	Censo				731
		Tlalixtac			
			Men	Women	
1882	Censo		217	239	456
1883	*Cuadros Sinópticos*		222	245	467
1900	Censo				461
1910	Censo				545
1921	Censo				745
1930	Censo				894
1940	Censo				921
1950	Censo				1,028
		Tonaltepec (Tanatepeque)			
arr. Span.	*Relación* (1580)	800 casas pobladas de gente			
1569	*Relación de los Obispados*	100 indios tributarios		(×2.8)	280
1580	*Relación de Guautla*	90 casados		(×3.3)	297
			Men	Women	
1882	Censo		59	64	123
1883	*Cuadros Sinópticos*		62	66	128
1900	Censo				136
1910	Censo				169
1921	Censo				182
1930	Censo				251
1940	Censo				273
1950	Censo				369
		Tutepetongo			
1544	*Libro de las Tasaciones*	Tribute in produce			
1548	*Libro de las Tasaciones*	120 pesos tribute per year			
1548	*Suma de Visitas*	66 cassas			

THE CUICATEC IN CULTURE CONTACT

1555	*Libro de las Tasaciones*	106 cassados 35 hijos y hijas de 12 años arriba	($\times 2 = 212$) $+35$ $\overline{247}$	($\times 3.3$) 350
		80 pesos, 109 fanegas de maiz (= 80 tributarios? 218 tributarios?)		($\times 1\%$) 274 264(?) 719(?)
1556	*Libro de las Tasaciones*	250 pesos de oro común		
1569	*Relación de los Obispados*	50 indios tributarios		($\times 2.8$) 140
			Men	Women
1882	Censo		109	112
1900	Censo			273
1910	Censo			341
1921	Censo			369
1930	Censo			370
1940	Censo			362
1950	Censo			431
1960	Censo		224	216	440

Ucila

1548	*Suma de Visitas*	938 "personas de todas hedades"		($\times 1\%$) 1,042
1560	*Tributos de Pueblos de Indios*	320 pesos, in money and cacao		

Zapotitlán

		Men	Women	
1882	Censo	66	66	132
1900	Censo			173
1910	Censo			183
1921	Censo			212
1930	Censo			153
1940	Censo			273
1950	Censo			290

equivalent terms, and should be multiplied by 3.3 to get total population. In later, post-1560s statements, when the tributary population was increased, while casados and vecinos are multiplied by 3.3, the tributarios are multiplied by 2.8. Finally, in the figures (such as in the *Suma de Visitas*) where one is given categories of "casados," "solteros," or "casados and muchachos de 15 años de abajo," I have multiplied the "casados" by 2.0 (for the wives), then added the "muchachos" or "solteros," then multiplied this figure by $^{10}/_9$ for the aforementioned children under three. And for the seventeenth- and eighteenth-century sources which give the data in "familias" I have (quite arbitrarily) multiplied these figures by 6.0.

I add a caveat. I do not believe that these numbers (2.8, 3.3, $^{10}/_9$, 6.0) represent any precise approximation, but rather only an order of magnitude. The arguments for deriving them are found in Cook and Borah (1960) and Borah and Cook (1960). However, it must be borne in mind while reading these that much of the calculation is based on not terribly reliable data, and on basic assumptions about sixteenth-century populations in central Mexico that have not yet been tested. As a result I do not insist very much on the exactness of the numbers I derive, except in that they represent an order of magnitude and a general trend. In Table 3 I have listed the raw statements from the sources together with the final estimate I have derived, so that anyone can rework the data if they have derived other factors for relating these figures to total population. I do not believe that these data can be pushed too far, and some of the calculations, particularly those of rates of decline and some of the extrapolations to pre-Hispanic population values, seen unjustified to me.

Similarly, I have not tried to derive population from the statements of amount of tribute. I am content to point out the general trend of declining tribute assessments over time in the *Libro de las Tasaciones* and to argue that it is a sign of declining population, without trying to derive mathematical indices of the rate of this decline. This is because I feel that too many other factors besides population affect these assessments, so that mathematical manipulations, especially with such a small sample as that which I have for the Cuicatec area, would be inappropriate.

The population data for the Cuicatec towns have been summarized in Table 3. There are certain points at which we can compile fairly complete population estimates for the Cuicatec region as a whole. The first these is 1548, from the *Suma de Visitas*, which is fairly complete for this area. The second is in the late 1560s, from the *Relación de los Obispados* (1904), generally dated to 1568, and from estimates from the *Libro de las Tasaciones* which supplement or reinforce the observations from the *Relación de los Obispados* and which range in date from 1564 to 1567. There is a jump of a whole century to the next fairly complete statement of population, from the eighteenth century. This is the *Theatro Americano*, which gives specific populations for the whole area. I have averaged in with these data statements by Thiéry de Menonville for 1777 and Ajofrín of 1766 (Thiéry de Menonville 1812; Ajofrín 1959). Then there is another gap in our information until 1882, when the *Cuadros Sinópticos* (Martínez Gracida 1883) were compiled. From then until the present we can construct a fairly complete population profile for the entire area from official censuses.

In order to generalize about population trends in the Cuicatec area from the above sources, while avoiding problems of spotty reporting, I have reduced my population profile to a restricted sample of towns within the Cuicatec area. They are chosen for two major criteria: they are Cuicatec, and they are represented in all of the above sources. Through time the units reporting the population have changed. However, it is possible to find evidence that covers roughly the same area, so that the figures from period to period are comparable. These towns cover both highland and lowland communities. I can describe the trend for all the Cuicatec towns and break this down into highland and lowland trends.

The towns covered in the sources vary in name over time. The area remained much the same, although there is a little slippage. Thus the *Suma de Visitas* figures include Coyula, although it does not appear in the later figures, and the 1568 figures included Nacaltepec, which is not in the *Suma* figures. Later, in the 1882-1960 figures, all of the Pápalos were included, while in some of the earlier sources these were part of either Pápalo or Tepeucila. In the lowlands, Guadalupe Obos, Guendalain (Valerio Trujano), and San Pedro Chicozapotes would originally have been part of Cuicatlán, while El Chilar would have been part of Alpizagua (Dominguillo). The figures are lumped

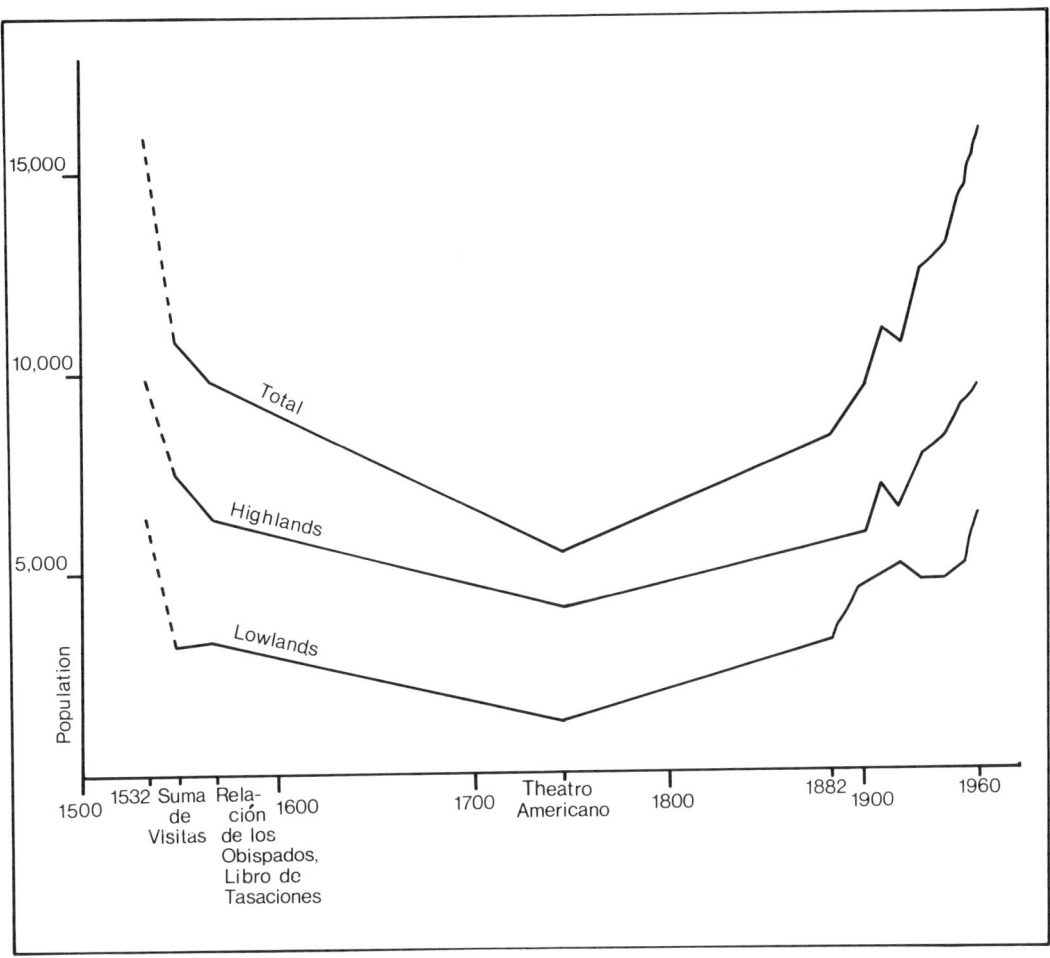

Figure 2. Population of the Cuicatec region from Spanish contact to the present.

together as highland or lowland areas. I am confident that these figures do in fact cover the same territory through time.

The towns used in this sample and the population figures for them at the relevant points in time are listed in Table 4. The trends that result from these figures can be seen in the graph of Figure 2. Table 3 shows population figures for all Cuicatec towns. It is easy to see that the sample in Table 4 is typical of the general trend.

The Cuicatec area followed the Mesoamerican pattern of great depopulation after the Spanish Conquest. There are some finer points one can deduce from this evidence, beyond the fact that great depopulation did take place. By the time of the first samples, from the *Suma de Visitas* in 1548, considerable depopulation had probably already taken place. There is no good direct evidence to tell what the extent of this depopulation was. There are a few statements in the *relaciones* (ca. 1580), but the estimates derived from these statements seem impossibly high. The *Relación de Atlatlauca* says that the population in 1580 was less than 5% of the population at the arrival of the Spaniards. Since they report 700 "indios cassados" at the time of the *relación*, this would give 14,000 "indios cassados" at the time of the Conquest, or 46,200 people for Atlatlauca. This seems an impossibly high figure, when compared to present-day figures. The *Relación de Tepeucila* states that before the Spaniards came, there were 40,000 indios tributarios. This would represent a total population of 112,000 people, which also seems impossibly high.

Universally, the highest population figures are the most recent ones we have. It is not clear that these represent the maximum possible population.

TABLE 3
POPULATION SAMPLE OF CUICATEC TOWNS

Town Name	Suma de Visitas (1548)	1560s	Theatro Americano (1745)	1882 Cuadros Sinópticos	1960
Highlands					
Cotahuixtla			168	169	578
Coyula	108		180		
Nacaltepec		560	198	275	536
Pápalo (Concepción)	1,664	1,904	1,188	288	972
Pápalo (Reyes)				1,218	820
Pápalo (San Andrés)				909	1,582
Pápalo (San Lorenzo)				341	283
Pápalo (Santa María)				361	560
Tepeucila	450	613	636	535	868
Teutila (San Pedro)	4,937	2,941	1,800	401	1,337
Teutila (Santa Cruz)				203	142
Tonaltepec		280		123	369
Tutepetongo	296	140		221	440
Total	7,455	6,438	4,170	5,814	9,427
Lowlands					
Alpizagua	63	84	216	351	385
Atlatlauca	2,171	2,454	408	998	849
Cuicatlán	1,053	840	750	1,255	2,456
El Chilar				195	834
Guendalain				500	1,216
Guadalupe Obos				113	134
San Pedro Chicozapotes				197	517
Total	3,287	3,378	1,374	3,609	6,391
Grand Total	10,742	9,816	5,544	9,423	15,818

The entire region is still essentially agricultural. It is not really significantly mechanized, except in transportation. The irrigation systems do not depend on any new techniques like motor-driven pumps, so the present population is probably less than or close to the population supported in pre-Hispanic times. Therefore, if we use the present population as an estimate of the pre-Conquest population, it will certainly not be too great, although it is possible that the pre-Conquest population may have been larger. So I can use the present population as a baseline to compare with earlier populations to see which areas were hardest hit and which were the quickest to recover from the depopulation.

When we compare the 1545 figures with those of 1568, we discover that the highland figures decrease, while the lowland ones stay substantially the same. And if we compare the 1568 figures to the 1960 figures, we see that the highland figures are 68% of the present population, while the lowlands are only 53%. This suggests that the lowlands were more quickly depopulated and that the depopulation of the highlands was slower and never as extreme as the lowlands.

The nadir of population in the graph for the Cuicatec region is the eighteenth century. Since we have no information for the seventeenth century, some recovery may have already taken place by the eighteenth century. At the same time, the area was clearly quite depopulated in the eighteenth century. The *Theatro Americano* (1952) contains, for example, a statement that Atlatlauca was so small and had so few people that it impoverished its *corregidores*. It could not afford its own government. Gerhard reports an injunction against the *corregidor* of Atlatlauca living in the town, because it had so few people and was so poor.

If we compare the eighteenth-century figures to the present population, the lowland towns sunk to 21% of the 1960 totals, while the highland figures only sank to 45% of their later population.

The different impact of depopulation on highland and lowland areas agrees with that reported by Cook and Borah (1960). Comparing the ratio of 1532 populations to 1568 populations, and highlands to lowlands, for all of central Mexico, they say:

> The mean ratio for the highland region is 5.42 ($^{1532}/_{1568}$) and that for the lowland regions, 14.80.... These data make it evident that there was a differential change in population, that of the coasts and lowlands tending to decline much more rapidly between 1532 and 1568 than that of the Plateau. [Cook and Borah 1960:41]
>
> There is, however, a very real difference between the highland and lowland: the rate of population decline was nearly twice as great in the lowland regions as in the highland ones ... in 1568 and 1580 the population in the lowlands was clearly declining at a much faster rate than that in the highlands—up to 1568, at a very much faster rate. After 1580 the decline was faster in the highlands, and slower in the lowland coastal areas, and by 1605, if we allow for probable bias in the data from the records of the civil congregation, the population in the lowlands was declining scarcely at all. Meanwhile, the rate of diminution on the plateau was still fast, although it, too, was abating. Our data indicate that demographically the lowland and coastal areas deteriorated faster and earlier than the plateau areas, but started to recover earlier. [ibid.:52]

While the Cuicatec data are far from adequate, it seems that this was also the case for the Cuicatec region. However, if we attempt to multiply the 1568 figures for the highlands and lowlands of the Cuicatec region by the ratios that Cook and Borah used for deriving pre-depopulation figures for 1532, we get numbers that seem far too high. While we confirm the trend they observe, we disagree numerically.

Certainly a large part of the population recovery has been quite recent. There is a noticeable rise of the population curve within the period for which we have good records, from 1882 to the present.

Spanish Introductions to Mesoamerica

The Spaniards introduced a number of new techniques, plants, and animals to Mesoamerica. Their assimilation followed the structure of the intercultural contact and the structure of the Cuicatec ecosystem. This included the effect of population change as it interacted with these structures.

Within the technological sphere there are three major elements that seem important. The first is iron, steel, and metallurgy in general. The second is the introduction of the whole complex related to draft animals. The third is the introduction of various new cultigens from the Old World.

At the time of the Spanish arrival, metallurgy in Mesoamerica was a recent arrival itself. This was nearly entirely confined to work with precious metals for objects of value. This aspect of metallurgy was intensified by the Spaniards, who were anxious to get as much gold and silver from the colony as they could. However, as there was only one small gold mine in Pápalo, which was covered early by an earthquake, this had little effect on the Cuicatec region. There was also some placer mining in the rivers, but that was also abandoned early.

The Spanish introduction of a relatively cheap kind of iron and steel was probably more important to Mesoamerica. Iron and steel tools replaced many of the wood and stone tools used previously and eliminated obsidian. Iron and steel seem to have been accepted immediately. I know of no good work on the implications of this change in Mesoamerica, which was undoubtedly major.

Large domestic animals were another important Spanish introduction. Their adoption illustrates the structure of the culture contact. Some could simply be added to the Indian system, filling niches that existed in the pre-Hispanic system. Others were compatible with the Indian social system, although they changed the economy somewhat. Others demanded or tended to be tied to the change to elements of the Spanish system.

The adoption of the Spanish cultigens was similarly affected by the structure of the Indian-Spanish opposition. Some of the cultigens were just added to the pre-Hispanic Indian inventory, or found in free substitution with other pre-existing cultigens. This is the case for many of the fruits. Others were complementary to pre-existing plants. However, other cultigens implied a complex of technical and social demands that tied them to the Spanish way of life. This seems to have been the case with sugar cane, just as it may have been the case with large-scale livestock ranching.

Livestock in the Cuicatec Region in the Sixteenth Century

Before the contact with the Europeans, the only important Mesoamerican domestic animals were the dog and the turkey, both of which were eaten. European domestic animals presented many new problems to the Indian economies. With domestic animals, the conflict between agriculture and pasture was introduced to Mesoamerica. Secondly, the introduction of domestic meat sources would result in changes in hunting patterns, as well as possible increases of meat in the diet. Thirdly, animals could be used for traction, both for plowing and perhaps more important, for transportation of products. Previously only manpower was available. This could be expected to have an impact on pre-Hispanic agricultural and transportation systems.

Finally, animals would be a source of manure for intensive agriculture, where previously the only significant manure was that from people themselves.

Chevalier (1956:77) reports that livestock very quickly spread into Nueva España. By the middle of the sixteenth century, livestock prices in Mexico were quite low due to the rapid expansion of livestock. Consequently one would expect that livestock would have been available for adoption in the Cuicatec area. The evidence of the *relaciones* shows that by 1580 livestock had been adopted in some towns and were used to some extent in all. However, at this time, when livestock had become quite cheap in all of Nueva España, it was not very important in any of the Cuicatec towns. Interestingly, livestock was adopted particularly in the lowland towns, while the *relaciones* of the highland towns state that because of the cold the highlands were unsuitable for livestock. Where livestock was found, in the lowland towns, it tended to be *ganado menor*, especially goats.

Thus, the *relaciones* say of Cuicatlán that goats were kept and were increasing rapidly, and of Atlatlauca that they kept all animals, though the Indians really did not go in for it much:

> Danse . . . en estremo las cabras que aquí se crian porque hay rramon verde todo el año con que se sustentan, y ase visto muchas cabras parir tres cabritos y criallos muy bien todos. [*Relación de Cuicatlán* 1955:188]
>
> De los traidos de España ay cauallos, mulas, cabras y obexas, aunque en estos pueblos no se dan los yndios a criarlos pero sabese que multeplica muy bien. Ay puercos y galgos y perros de que los yndios ya se siruen para caçar. Ay gallinas . . . de Castilla [*Relación de Atlatlauca y Malinaltepec* 1905:174-75]

The *Relación de Atlatlauca* also contains a statement that "el dia de oy . . . tanbien comen carneros, obexas quando las alcançan . . ." (*Relación de Atlatlauca y Malinaltepec* 1905:171). This suggests that perhaps livestock was not that common in the region, so that not everybody could always afford to eat domestic meat. This is in contrast to the situation described by Chevalier for the rest of Mexico, or at least the parts heavily occupied by Spaniards.

In the highland communities livestock was not usually considered to be viable. The *Relación de Pápalo* tells us that "Ganados de cabras ni obejas no se da por la frialdad y asperaza de la tierra" (*Relación de Papaloticpac* 1905:92). The *Relación de Tonaltepec* says, "hay pocos pastos, é los que hay no son provechosos á los ganados menores" (*Relación de Guautla* 1962:14). The only introduced animals for which there is evidence are "aves de Castilla" in Tonaltepec (*Relación de Guautla* 1962:16) which could probably have been cared for in the same way as the "aves de la tierra," that is to say, turkeys. Tepeucila is also said to have had "gallinas castellanas" (*Relación de Tepeucila* 1905:98). The people of Tepeucila also ate "otras carnes de nuestro ganado" (*Relación de Tepeucila* 1905:96), in addition to the meat of the deer, which was expensive. It is not clear if domestic meat also cost "excesivos precios," but if it did, meat cost considerably more than it did in other parts of Mesoamerica where the Spaniards had livestock haciendas.

Almost at the end of the sixteenth century, we have evidence of *estancias de ganado menor* granted to two Cuicatec towns, Nacatepec (Nacaltepec) in 1590 (Spores and Saldaña 1973: No. 974), and in 1597 two *estancias de ganado menor* were granted to two different individuals in Cuicatlán (Spores and Saldaña 1973: Nos. 278, 279).

Thus we see that, at least in the sixteenth century, the impact of Spanish livestock was limited, especially on Cuicatlán. Significantly, the "ganado," where they were adopted, were goats or pigs and chickens. The pigs and chickens could fit into the same pattern as turkeys and dogs in the pre-Hispanic pattern, and would simply have been raised by individual families.

The goats present a slightly different aspect. They definitely constituted an innovation. However, they may not have competed directly for land that could be farmed, in the Cañada towns. The reference to goat herding in the *Relación de Cuicatlán* tells us that they could grow well because there was "rramon verde." Goats can live off the xerophytic scrub forest that is the natural vegetation of the low Cañada, utilizing a part of the environment (the unirrigated part of the Cañada) that previously would only have been used for hunting and firewood. Presumably domesticated animals would be more efficient as a source of meat than hunting. At the same time the dropping population would have relieved the pressure on the land, so that in the period of the introduction there would have been more room for goats.

The sixteenth-century evidence that we have for adoption of livestock follows the Spanish division of *ganado menor* and *ganado mayor*. *Ganado menor* were added to the Indian communities either in niches that already existed (as with pigs and chickens) or that could be worked in, in a complementary way to the pre-existing system, as in the case of goats. In part, too, *ganado menor* represent much smaller capital investments then *ganado mayor*.

In contrast, *ganado mayor*, specifically cattle and horses, had definite implications that tied them to Spanish rather than Indian systems. As traction animals, for transportation along the Camino Real, they facilitated long-distance travel and trade, which as we have seen, the Spaniards took over early. They made long-distance trade possible over a longer distance, and in greater bulk. The change from human to quadruped traffic on the Camino through the Cañada demanded some changes in the trail itself. Spores and Saldaña list a document (1973: No. 1108) that orders that the roads to Mexico and Guatemala be straightened in 1560.

> 1108. 1560 OAXACA. Caminos. Comisión a Cristóbal de Espíndola, alcalde mayor de esta ciudad, para que vea se aderecen los caminos que salen a Guatemala y México, según la relación que de ellos hizo Juan Gallego, procurador de dicha ciudad. Vol. 5, Fa. 83.

Large-scale cattle herding is not compatible with the agriculture systems of the Cañada. Cattle cannot graze as successfully as goats on the xerophytic scrub of the Cañada. Their herding demands wide spaces and a hierarchical kind of hacienda specialized in their production. Finally, horses were status items among the Spaniards and were tied to Spanish privileges.

One can treat cochineal and silk as introduced animals. Silkworms were introduced by the Spaniards, and, while cochineal (tiny insects that lived on nopales and produce a bright red dye) were from Mesoamerica, the demand for them was increased greatly by the Spanish ties to the world market (Hamnett 1971; Borah 1943). Both sericulture and cochineal were found in the highlands rather than the lowlands. The impact of silk on the Cuicatec area was not as great as on the Mixtec area, but cochineal was important. The Cuicatec towns for which they were most important were the towns under the Mixtec sphere of influence.

Pápalo had,

... morales con que se cria seda, avnque por la frialdad de la tierra y ser tan ayroso este pueblo se da mal la seda en el, y ansi se cria poco. [*Relación de Papaloticpac* 1905:92]

Tepeucila also had a little cochineal.

En algunas estancias en este dicho pueblos se da alguna grana, porque como es serrania no tienen espaçio ni lugares descanpados para podello senbrar y ansi es poco lo que cogen. [*Relación de Tepeucila* 1905:97]

In contrast, Tutepetongo, which was tributary to the Mixteca, had "morales para criar seda . . . é tunales para criar grana . . ." and "cogese seda en este pueblos" (*Relación de Guautla* 1962:13). And in Tonaltepec, also under the sway of the Mixteca, "dase seda y la crian los naturales, hay árboles do crían é cogen la grana, é dicen que no se da." Also, "tiene por trato criar seda" (*Relación de Guautla* 1962:16). However, one still does not get the impression that they had a monoculture of silk or cochineal like that characteristic of some of the Mixtec towns later.

The importance of both these products was that they were demanded by the world market outside the Cuicatec area, indeed, even outside Mesoamerica. This did not mean that the towns that produced silk and cochineal were forced into the Spanish pattern of life. The exploitation of these crops was quite compatible with the Indian, symmetrical, egalitarian life. However, the complementary relationships with the Spanish hierarchy above the towns often meant that these towns had to produce these products for their tribute or taxes. This reminds one of the pre-Hispanic demands of the Aztec tribute system that forced highland Indians to rent lands or work in the lowlands.

The structural forces exerted by the trade in grana and silk instead affected the lowland communities, because it was through them that the grana and cochineal were funneled on their way on the Camino Real to the outside market. It was in towns like Cuicatlán that the first Spaniards lived, collecting the grana for export. This force for schismogenesis influenced the lowland towns towards the Spanish end of the cultural opposition.

Introduced Plants

Cortés, in his letter of 15 October 1524 from Mexico to the King of Spain, advises the importation of European fruits and plants.

Tambien he hecho saber á v. Ces. M. la necesidad que hay que á esta tierra se traigan plantas de todas suertes, y por el aparejo que en esta tierra hay de todo género de agricultura; y porque hasta ahora ninguna cosa se ha proveido, torno á suplicar á B.M, porque dellos será muy servido, mande enviar su provision á la casa de la Contratación de Sevilla para que cada navio traiga cierta cantidad de plantas, y que no pueda salir sin ellas, porque será mucha causa para la población y perpetuación della. [Cortés 1886:322]

At first, with a few exceptions, most of the crops introduced by the Spaniards were temperate and were accepted differentially along the lines of the highland/lowland break of the Cuicatec communities. In the highlands we are told that many new fruits and vegetables grow well and are grown, while in the lowlands there are often specific statements that crops from Spain would not grow in *tierra caliente*.

The *Suma de Visitas* (1905:186) tells us that in Pápalo in 1549, "tiene . . . muchos morales y frutas de Castilla. . . ." The *Relación de Pápalo* for 1580 lists

. . . arboles de Castilla que se an plantado despues que vinieron los españoles como son duraznos en cantidad, perales y algunos menbrillos y mançanos y granados y algunas higueras avnquel fruto dellas es desabrido por la mucha humedad que partiçipan. [*Relación de Papaloticpac* 1905:92]

It also says that "Dase bien qualquier hortaliza ansi de lechugas como de rravanos y coles . . ." (ibid.).

The *Relación de Tepeucila* says that

de Castilla muchos duraznos que se dan bien, y perales que ynxieren en los mançanos de la tierra y ansi la fruta es menuda y de ruyn sabor. . . .

Dase rrazonablemente la ortaliza de lechugas, rravanos, coles, y cebollas y otras legumbres que las semillas se an traydo de Castilla. . . . [*Relación de Tepeucila* 1905:97]

In contrast the lowland communities seem to have had much less success with fruits and vegetables from Spain. The *Relación de Cuicatlán* (1905:188) says that "acausa de ser la tierra tan calida no se da ninguna semilla de castilla, ni hortaliza, avnque se cojen rrazonables melones." This blanket denial is modified a little by statements that there are "muchos naranjos de castilla." Similarly, the *Relación de Atlatlauca y Malinaltepec* 1905:173) tells us that "no se da fruta de España mas de las dichas, porque, como es tierra caliente, no se da. . . ." The exceptions mentioned are oranges and "cañas duzes como las de España de que se haze el açucar" (ibid.).

The *Suma de Visitas* says of Alpizagua (Dominguillo), a *tierra caliente* town, "dase trigo y frutas de castilla dos o tres vezes en el año." This is exceptional, both for a lowland town, and for any town in the Cuicatec area at this time. It may represent an early Spanish settlement that did not continue, since later Dominguillo was typically Indian.

In the highlands, plums, apricots, pears, apples, pomegranates, and even figs, as well as vegetables like lettuce, radishes, and cauliflower were grown. In the lowlands, the only two significant importations seem to be oranges and sugar cane.

The plants that were not adopted are as important as those that were. The *relaciones* are also full of statements, sometimes somewhat querulous, that various plants could be grown but the Indians would not plant them. From the highland town of Pápalo the *relación* says

> es tierra aparejada para darse bien en ella trigo avnque no lo sienbran, porque esta desviada de poblazon de españoles y los naturales no se saben aprovechar dello. [*Relación de Papaloticpac* 1905:92]

The *Relación de Tepeucila* (1905:97), also in the highlands, follows its statement that various Spanish vegetables grew reasonably well, with the proviso that "los naturales no se dan mucho por este, mas que por mayz y frisoles y chile y otras legumbres que ellos llaman «quelites» que todo esto se da en esta pueblo."

From the lowlands the *Relación de Atlatlauca y Malinaltepec* (1905:173) tells us that because it is *tierra caliente*, fruit from Spanish sources does not grow that well, "etceto melones que si se diesen a senbrarlos se darian," and then goes on to explain that

> Alguna ortaliza de España se diera si la senbraran, como coles y lechugas, perexil y culantro, pero los yndios no lo sienbran ni se dan a ello por ser mas afiçionados al maiz. [*Relación de Atlatlauca y Malinaltepec* 1905:173-74]

Thus it seems that some fruits from Spain got a fairly firm grip on the Indian towns in the Cuicatec region quite early, but others, particularly wheat, were not adopted, probably in spite of considerable pressure for their adoption. However, the pressure for wheat came from populations of Spaniards. As the *Relación de Papaloticpac* pointed out, there was no Spanish population in the area at the time of the sixteenth century except the few Spaniards in Cuicatlán mentioned by Padre Ponce.

The adoption of the vegetables followed the Spanish/Indian division. Since the Cuicatec area was abandoned early by Spaniards, except for the most minimal administrative and religious services, the vegetables and fruit adopted were those that were compatible with the Indian culture as it was evolving, in opposition to the Spanish one. In contrast, wheat was identified strongly with the Spaniards, and as such was resisted by the Indians. It was not eaten by the Indians, and not grown by them unless there was a large population of Spaniards nearby. Its presence in the agriculture is symptomatic of the presence of a Spanish market.

Summary: The Impact of the Spanish Conquest

Looking back from the end of the sixteenth century, on the impact of the Spaniards on the Cuicatec area, one's first impression is that the immediate effect was not that great. The Spaniards with a great flourish divided up the area, as they did the rest of Mesoamerica, into administrative districts, both secular and religious. However, none of the towns in the Cuicatec area turned out to be a particular prize. They were too small, had too few Indians per unit, had no special resources like gold mines, and were too far from the main centers of population to be very interesting. As a result they quickly dropped from the encomendero to the royal crown. One encomendero even left for the Philippines in the hopes of doing better.

One does not get the feeling that the *corregidores* fought for the Cuicatec *corregimientos* either. Gerhard cites an order that the *corregidor* of Atlatlauca not live in the town because it was too small to support him. This seems to have been the tone of all the Cuicatec towns. Similarly, the only ambitious religious establishment in the area was the convent for some 20 friars in Teutila, which was abandoned within 30 years of its establishment. The Cañada and the Cuicatec towns were left with minor petty administrators and some hard-working priests who had to learn the local language to administer the doctrine to a flock scattered through the mountainous Cuicatec region.

The change in rulers and even the religious change did not seem to have had a great effect on the Cuicatec ecosystem. What is a major change is the new importance of the opposition between the local, Indian system, and the national, Spanish system. This division in part was anticipated by the Aztec-Cuicatec relationships, but Spanish contact resulted in an intensification and restructuring of this hierarchical relation. In the sixteenth century, the Cuicatec region was almost entirely on the Indian side of the opposition.

This fact influenced all of the Spanish elements that were accepted and incorporated into the Cuicatec ecosystem. The plants and animals adopted fit what were to become Indian complexes throughout the history of Mexico. The only areas within the Cuicatec region that reflected in microcosm the hierarchical Spanish system were the larger administrative centers, especially those that were on the Camino Real. Thus Teutila (which was not on the Camino Real) retained its administration for a while and had Spanish administrators resident in it. Cuicatlán always had a small Spanish community of priests, administrators, and merchants. Various priests lived and traveled alone through the other Cuicatec towns, administering their scattered parishioners in the native languages.

The end of the sixteenth century saw most of the elements present that were to operate in the processes of the next 400 years. These are the natural system and the basic agricultural adaptations made to it by the pre-Hispanic populations. Added to this is the opposition between the Indian and Spanish sides of the culture contact, as well as a number of elements introduced by the Spaniards. By the end of the sixteenth century, the structures of the two were established, and the new and old elements had adapted to fit these structures. All of these interactions of the two cultures were influenced by the changing population and the technological effects of the larger world in which they were set.

Chapter 4
The Cuicatec Ecosystem from the Conquest to the Present

Up to now we have emphasized that the Spanish Conquest took place fairly recently. Because of this we have good descriptions of the Cuicatec ecosystem at the time of the Conquest. However, the Spanish Conquest did take place more than 400 years ago. This is more than enough time to be able to observe processes of change over time operating in the ecosystem. This chapter traces the history of the Cuicatec ecosystem over these 400 years and analyzes the historical process that affected the ecosystem over this period.

The last chapter described the form of the Cuicatec ecosystem at the end of the sixteenth century after the initiation of contact between the Spanish and Indian cultures. This culture contact continued through the history of the ecosystem. As various factors interacted in the ecosystem, one can see the ecosystem change.

The Seventeenth Century: The Introduction of Sugar Cane

The only major source that covers the seventeenth-century Cuicatec is a compilation of the titles of documents relevant to the study of Oaxaca from the Ramo de Mercedes of the Archivo General de la Nación, in Mexico City (Spores and Saldaña 1973). The seventeenth century was an important period, for during this period the population of the Cuicatec area probably reached its nadir. Unfortunately, no source from the seventeenth century gives direct evidence on population. There is some evidence of some other processes which had begun in the sixteenth century.

In the sixteenth century, some European crops had not been adopted in the Cuicatec region. While we have a reference to "cañas duzes" in the *Relación de Atlatlauca*, sugar became important in the Cañada during the next century, the seventeenth. The first reference we have for a *trapiche* (sugar mill) and "tierras y aguas" to grow sugar, is in 1643, for Cuicatlán (Spores and Saldaña 1973:No. 2833). In 1664 another trapiche got a license to make sugar, as well as permission to plant sugar cane. The trapiche made sugar "valiendose de los arroyos comarcanos" (1973:Nos. 284, 285). Depopulation had already left its mark. The pressure on the irrigation system must have considerably diminished, if it was possible to divert part of the water from the irrigation system to run a trapiche. The trapiche established in 1664 was still in operation in 1675 and in the hands of the same man, Francisco Fernández Machuca, "vecino de la Ciudad de Antequera," who had a "hacienda y trapiche que posee en esta jurisdicción" (1973:No. 286). In 1675 there was also at least one other trapiche in Cuicatlán, in the hands of a *clerigo presbítero*. At least some of the Spaniards who settled in Cuicatlán as administrators were taking advantage of sugar growing to make a profit.

The only other references to a trapiche in the Cañada was in San Blas, between the towns of Atlatlauca and Guaxolotitlán (Huitzo) nearer to Atlatlauca. This also belonged to a *clerigo presbítero*. As we have noted with the *estancias de ganado menor*, this was on the Camino Real, at the end of the Cañada nearest the Etla arm of the Valley of Oaxaca.

The commercial production of sugar cane implies certain conditions. Sugar cane demands frost-free climate, so it was only found in the lowland communities. The *Relación de Cuicatlán* does not mention sugar, but it is mentioned for Atlatlauca. Part of Tutepetongo must have been low enough to grow sugar cane, since its *relación* also lists *cañas de azúcar*.

More important, commercial production of sugar cane demanded a complementary, hierarchical, and hence Spanish system. It could not be done within the symmetrical structure of an Indian village. François Chevalier compares the demands of sugar with those of wheat:

> pero, al paso que el trigo se cultivaba sin grandes costos en terrenos trabajados por poca gente, los ingenios azucareros de la Nueva España—como los de las Antillas—fueron al contrario, desde un principio, grandes explotaciones mitad agrícolas, mitad industriales, que empleaban centenares de hombres, indios o negros . . . si se exceptúan los molinos movidos a mano y otras explotaciones familiares, que aún subsisten en nuestros dias, el más modesto trapiche constituye una verdadera empresa en pequeño, que requiere por lo menos una quincena de hombres trabajando de manera continua durante algunos meses después de la cosecha. Además, el cultivo de la caña es exigente e intensivo; necesita muchos cuidados, labores profundas, trabajos de riego, esto es, una mano de obra abundante, animales y herramientas. En otras palabras, los ingenios no podían encontrarse más que en manos de capitalistas, grandes o chicos. [Chevalier 1956:63]

These demands could with difficulty be met by the largely self-sufficient and symmetrical structure of the Indian villages. The Indians were protected by the royal laws from exploitation in this manner. Even when these protections broke down, as they frequently did, the demands of the sugar ingenio or trapiche forced a reorganization that inevitably led to a replacement of the Indian community with a different kind of community. Frequently, rather than using Indian labor, which was protected, Negro slaves were imported. Chevalier sketches the reasons for this:

> El dueño de un ingenio azucarero tenía que disponer de tierras fértiles, de agua abundante y de un equipo costoso. Pero el problema esencial era conseguir mano de obra apropiada. En efecto, el cultivo de la caña es delicado y trabajoso, y, sobre todo, las explotaciones azucareras exigían trabajos a los cuales difícilmente resistían los indígenas, porque eran demasiado duros o porque requerían un esfuerzo sostenido. Los hombres que se ocupaban de las calderas, de las prensas, y en general de la fabricación del azúcar, tenían que ser por fuerza negros, como lo demostraba la experiencia. Por añadidura, estaba prohibido bajo penas severas emplear indios en esas tareas. [Chevalier 1956:221]

Using Bateson's model of the process of schismogenesis between the Indian and the Spanish pattern, we observed that Indians were essentially egalitarian, symmetrical, peasant communities. But sugar cane demanded a hierarchical, complementary organization for its exploitation. Therefore sugar cane demanded that a community exploiting it change to the Spanish structure.

This meant that with the growth of importance of sugar cane within the Cañada from the seventeenth century on, for the first time the schismogenesis between the Spanish and Indian communities existed on the local level. In the sixteenth century the Cuicatec communities were still completely Indian. The Cuicatec ecosystem was symmetrical, as opposed to its complement, the hierarchical Spanish system. However, now, in addition to the symmetrical relations within the Indian communities and symmetrical relationships between the Indian communities, there were complementary relationships between the Indian communities and the sugar haciendas, and a complementary structure within each hacienda. This could affect only the lowland Cañada communities, where sugar cane could grow. In the highlands, the symmetrical Indian communities continued.

Sugar cane cultivation, like the commercial demands of grana, was another force towards the Spanish side of the schismogenesis that affected the lowland communities. Through the eighteenth century it grew in importance. At the beginning of the seventeenth century, the entire Cuicatec area was still mostly Indian. However, while the Cañada began the seventeenth century homogenous and Indian, it entered the eighteenth century with the seeds of schismogenesis planted between the Indian and Spanish patterns.

The Eighteenth Century: The Spaniards Establish a Foothold

There are several sources for the eighteenth century. The first of these is the *Theatro Americano*, which is a compilation based on geographical descriptions collected after a *cedula* of 1743. This was condensed by the cosmographer of Mexico, José Antonio Villaseñor y Sánchez, and published in summary form in 1745. Gerhard (1972:32) warns that it "contains many topographical and factual errors and serious omissions." Certainly, in many cases the linguistic ascriptions in the *Theatro* seem fairly capricious. Still, it is the first major source after the seventeenth century, and it is fairly comprehensive for the Cuicatec area.

Most of the population data for this period come from this source. The population figures in the *Theatro Americano* were lower than the last figures from the sixteenth century. From this point on, the population rose. Because of this, we suggested that the population reached its nadir sometime in the preceding century and had started upward by the eighteenth century.

The Cuicatec region was still apparently mostly Indian in 1745. There were few Spaniards in Cuicatec towns. The *Theatro* points out that one Cuicatec town, Atlatlauca, had so few people and was so poor, that it impoverished the Spanish administrators sent there.

While the sixteenth-century Cañada towns were intensive specialists in "cash cropping" of food, agriculture in the Cañada in the eighteenth century seemed much less specialized. The *Theatro* described all of the lowland towns as producing maize for subsistence, as well as some fruit. There is no mention of any extensive commerce in any of these foodstuffs.

In contrast the highland towns had two specialized activities, both originally pre-Hispanic, but both tied to demands from the Spanish system. These are weaving of cotton and grana, or cochineal, production. There is also one town (Cotahuixtla) that followed the pre-Hispanic specialization of making petates from palma, which they still do.

The *Theatro Americano* listed cotton weaving as the *granjería* (trade, occupation) of Tepeucila and of Teponaxtla. Ajofrín, in 1766, noted that in the towns around Teutila, the Indian women wove cotton, while the Indian men went out to the other towns and provinces trading for cotton to weave and selling woven cotton, just as they had in the sixteenth century.

Grana, too, continued to be important throughout the centuries in the highland towns. The *Theatro* mentioned grana in 1745 for Coyula, Tepeucila, and Teponaxtla. Ajofrín mentioned there were Spaniards in Cuicatlán engaged in the cochineal trade. The eighteenth-century *relación geográfica* of Pápalo, in 1778, included a lengthy description of the care of the grana in Pápalo, including a list of its enemies, and the fact that they had three crops a year, one of which was grown in the lowlands. The grana for the cold-season crops was reintroduced every year to the highlands. Grana and cotton could be produced, since they were pre-Hispanic in origin, within the symmetrical Indian communities. This production was sanctioned and encouraged by the Spanish hierarchy, who handled its marketing to the Spanish macrosystem and to the world. The route through which these products entered the Spanish macrosystem was the Camino Real. Because the lowland Cuicatec towns were the only ones of the ecosystem on this route, they were on the interface between the Spanish system and the Indian one. Ajofrín tells us that the merchants who dealt in grana lived in the lowland towns in the Camino Real.

The *Theatro* says of Tehuacán, "y por estar en el camino real, que entrra a las provincias de Oaxaca, y goathemala no carece de trafico, ni de passageros su comercio" (*Theatro Americano* 1952, II:349).

This also applied to the Cuicatec Cañada towns on the Camino Real. Thiéry de Menonville was able to rent horses in 1777 in relays all through the Cañada through an officially regulated system of "topiths" who had an official scale of pay and who guided travelers from one town to the next.

The Camino Real put the Cañada towns on the interface between the Spanish system and the Indian system. As we have seen, the relations between the Indian system and the Spanish were always hierarchical, with the Spaniards on the top and the Indians on the bottom of the hierarchy. Because the interface between the systems was in the lowland Cañada towns, they in turn began to have a complementary hierarchical structure within the individual towns. The road, like sugar cane, was a force for schismogenesis from the Indian pattern towards the Spanish pattern. Through the eighteenth century we can see the process of this schismogenesis.

The Cañada at the start of the eighteenth century was almost entirely Indian, and was engaged in subsistence agriculture. The *Theatro Americano* did not mention specialized Spanish crops nor any significant sugar production, although there must have been some small-scale production. It also did not mention either Spaniards or Negros, who were linked to sugar production, nor the whole specialized complex of commercial sugar production.

However, in the earliest source for the eighteenth century there was to the north of the Cañada, in the southern part of the Valley of Tehuacán, a finger of sugar, Spaniards, Negros, sugar mills, and even wheat. As the century progressed, this finger moved into the Cañada.

In the *Theatro* of 1745, there were españoles, mestizos, and mulatos living in Theutitlán del Camino, and an ingenio near Los Cues worked by Black slaves. By 1766, Ajofrín noted wheat (implying a Spanish market) at Venta Salada, and a "trapiche de azúcar" (in the hands of Jesuits) at Los Cues, in the lower Valley of Tehuacán. He also noted some trapiches on the road between Cuicatlán and Dominguillo and at Atlatlauca.

By 1777 Thiéry de Menonville saw wheat and a sugar plant worked by Negros at Los Cues. He observes that Blacks were necessary because Indians could only (by law) be engaged for 1 month to 40 days and therefore were always coming and going, so they did not learn the necessary skills and also could not be counted on when needed (Thiéry de Menonville 1812). He also saw wheat growing in Cuicatlán. The *relación geográfica* of Cuicatlán for the same year tells us that there were trapiches at Thecomaslagua and at Guendalain across the river from Cuicatlán, as well as at Collula (Coyula).

This crop invasion was paralleled by an increase in Spanish population, or at least of non-Indians. The *Theatro Americano* reports the entire area to be Indian, except, one supposes, the *alcaldes mayores* and the priests. But later, in 1766, Ajofrín noted that

> La jurisdicción [de Teutila] se compone de . . . todos indios, pues, como la tierra es tan escabrósa y áspera, no se han radicado familias de españoles, que sólo buscan el interés en minas o haciendas. Lo mismo sucede en todo el obispado de Oaxaca, a excepción de algunos pueblos situados en caminos reales, que se encuentra gente de razon, asi llaman a los que no son indios—y españoles, y por eso toda esta tierra poseen los indios. [Ajofrín 1959:62]

He also tells us of Cuicatlán that,

> Está en el camino real para Oaxaca, Guatemala y otras provincias, y por eso hay algunas familias de Españoles con tiendas y comercio de grana, pero la mayor vecindad se compone de Indios cuicatecos. [ibid.:82–83]

Still later, in 1777, Thiéry de Menonville remarked that Cuicatlán was more civilized than the towns around it. He mentioned the cultivation of wheat and the pruning and grafting of trees as signs of this civilization. He also described the priest rehearsing motets of his own composition in the church of Cuicatlán.

The eighteenth century was a period when the equilibrium between the Spanish and Indian culture patterns was changing. This change happened in the lowland communities more than in the highland communities. This was in part because of the differential impact of the depopulation on the lowland and highland adaptations. It was also in part due to several characteristics of the lowland communities not present in the highland communities which made them vulnerable to forces that pushed them toward the Spanish end of the schismogenesis.

The eighteenth century represented a change in the equilibrium because it is in the eighteenth century the population began its recovery from the massive depopulation that followed the Spanish Conquest. This depopulation was more severe in the lowlands than in the highlands. It also had a different effect on the lowland Cañada towns than on the highlands, because of differences in their adaptations.

In the highland towns in the eighteenth century lived subsistence agriculturalists who paid their taxes and were tied to the national economy by producing woven cotton products and cochineal. Both these products were pre-Hispanic in origin and were fully compatible with the post-Hispanic, symmetrical community structure. This symmetry was in contrast to the complementary hierarchical relations with the Spanish market for these products.

The lowland adaptation was originally a kind of specialty food production. The production depended on a high overall population that made intensive food production worthwhile. The massive depopulation meant that the highland towns that had traded for the Cañada food products suddenly had much less pressure on the land. As a result they did not need to buy products from the Cañada irrigation systems. At the same time the population of the lowlands declined more than that of the highland towns, resulting in abandonment of some of the irrigated areas and a general shift from intensive specialty production of crops to a more subsistence-oriented system. This unintensive kind of lowland community was then subjected to a number of influences that tended to upset the equilibrium of the symmetrical Indian communities and change them to the complementary Spanish structure.

One of these was the increasing competition of sugar cane production with lowland agriculture. At first glance this seems to be another kind of specialized food production. However, sugar cane demanded a Spanish form of complementary structure for the communities that depended on its exploitation. Further, sugar cane production often implied the introduction of non-Indian populations into the region. The haciendas that began to fill in the spaces in the depopulated Cañada were often dependent on Negro slave (and later "free") labor for sugar cane production.

As a result of this, the Cañada of the Cuicatec

ecosystem began to have Spanish and Indian communities existing side by side. At the same time there was another force that tended to make the Cañada communities move towards a complementary organization within each town. This was the effect of being located on the Camino Real, the major route of north-south communication in Mesoamerica and the primary link between the Cuicatec ecosystem and the macrosystem to which it belonged.

Because of the Camino Real, the Spanish administrators, merchants, and priests moved through the system and established their bases in the Cuicatec towns on the Camino Real. Because of this, the interface of Spanish-Indian relations took place in the Cañada towns. As we have seen, this interface was always hierarchical, so to the extent that these relations took place in the lowland Cañada towns, these towns tended to become increasingly complementary in their structure, and therefore to move towards the Spanish pattern.

The Nineteenth Century: Roads, Railroads, and Haciendas

The eighteenth century ended with increasing forces for schismogenesis working on the Cañada towns of the Cuicatec ecosystem. These pressured the Cañada towns toward the Spanish structure. In the nineteenth century another change increased the pressure on the Cañada towns while hardly affecting the highland towns. This was the pressure caused by significant improvements on the communication route through the Cañada.

Earlier I suggested that one of the impacts of the introduction of draft animals to the Indian system would have been to change the patterns of transportation by allowing products to be transported farther and more cheaply. This would have been a quantitative change in the patterns. There is some evidence that the pre-Hispanic route had to be changed somewhat to accommodate draft animals. However, the basic pattern would remain the same as it was with human porters in pre-Hispanic times, because the Camino Real was only passable to pack animals.

Through the eighteenth century, the Camino Real was only passable by pack animals or on foot according to the testimony of travelers, from Padre Ponce to Padre Ajofrín, regarding the narrowness and steepness of the trails. Thiéry de Menonville, who had an eye for the military logistics of the route, states specifically that the Camino Real was not passable to wheeled traffic when he traversed it in 1777:

> When at dinner, at an after period, with the intendant of St. Domingo, on his asking me respecting the roads in Mexico . . . I therefore merely answered his interrogation by telling him in general terms, that I found them very bad; and in good truth, though the road I was now travelling was that of Guatimala, and the only highway on which is transported the various produce of a valley, which extends four hundred and eighty leagues, I did not find thirty leagues of road on which a carriage could pass. [Thiéry de Menonville 1812:831]

However, in the nineteenth century, steps were taken to improve the road, culminating when Juárez pushed a road through the Cañada. This road was begun in 1833 and 1844, but was finally constructed from 1850–65 (Iturribarría 1955:159; Esteva 1913:97).

The construction of a road open to wheeled vehicles and regular wheeled traffic had a major effect in changing the relationship of the Cuicatec ecosystem to the rest of Mesoamerica. It opened up the area to which produce of the ecosystem could be transported. This benefit would not be uniformly extended to the entire ecosystem. Instead, the disparity in ease of travel between the Cañada and the highlands would be increased. The highland towns had been Indian in part because of their isolation from the intrusion of the Spanish macrosystem. Since this improved road followed the all-weather road that went into the mountains from the Cañada towards Oaxaca at Dominguillo, Atlatlauca, which previously was on one of the two routes that pack animals would follow, would now have been isolated from traffic on the national artery. From this point on, to the extent that the forces of schismogenesis affecting the lowland Cañada towns depended on communication with the macrosystem, Atlatlauca would follow the same pattern as the highland Cuicatec towns, tending to remain Indian.

Almost all of these effects of the improvement on the road would have reinforced the complementary Spanish pattern. The road increased ease of access of the national system to the towns it touched. The national system, as we have seen, was solidly in the Spanish pattern. Secondly, the improved road would increase the tendency of highland products to be funneled down through Cañada towns on the road. This in turn reinforced the formation of a complementary structure in the towns along the road between the Spaniards and the Indians in the Cañada towns. Thirdly, it became more economical for products of the Cañada towns to be cash cropped and shipped out to the national economy. This was particularly advantageous to the sugar haciendas established in the Cañada.

The road improvement made from 1850–65 established a qualitative change in the importance in the Cuicatec ecosystem as to differential access to communication with the rest of Mesoamerica. This pattern was intensified along the same lines by the establishment of a narrow-gauge railroad in 1892. The railroad left the Cañada in the middle, ascending the canyon of the Tomellín (Fig. 1). This did not cause Dominguillo to be "hung out to dry" as Atlatlauca had been earlier, because Dominguillo had good access to the railroad along the floor of the Cañada on the road that was already built.

The road was a force for the schismogenesis operating on the Cañada part of the Cuicatec ecosystem. It worked on a pattern that had its roots in the eighteenth century. In the nineteenth century we see this schismogenesis start to produce a definite pattern in the Cañada towns. This was the formation of two kinds of settlements in the Cañada. One was the hacienda, producing sugar cane, which, as we have seen, was complementary and Spanish in structure; the other was the Indian town with a subsistence and mixed-fruit economy which was increasingly beleaguered by forces for schismogenesis from which the highland towns remained free.

The sugar haciendas, as we have seen, were not compatible with the Indian structure. They also usually represented not only the importation of overlords and the imposition of a new social structure, but also the importation of a new population of workers, Negro slaves. This was possible because at the time of the nadir of population, there was room in the interstices between the Indian communities in the Cañada for the haciendas to spring up.

Esteva (1913) mentions the Cañada specifically when he tells us that slaves were freed in 1825. This shows that they were used on the sugar ingenios that began to become important in the eighteenth century. By the end of the nineteenth century the haciendas were well established and expanding their irrigation systems. The *Cuadros Sinópticos* (Martínez Gracida 1883) tell us of three important haciendas in the Cañada. One is the hacienda of Guendalain, which existed from the sixteenth century (Hamnett 1971). It is said to have had three aqueducts, one constructed in 1700, and the other two constructed in 1843 and 1871, respectively.

Another sugar cane hacienda was Tecomaxtlahuaca, which is south of San Pedro Chicozapotes on the east bank of the Río Grande on the route to the present dam of the Matamba Canal. Its abandoned ruins still stand. This hacienda was reported to have two aqueducts, one from "1700," but the other from 1818. This hacienda was supposedly founded as an ingenio in 1540. Finally, Guadalupe Obos, just to the north of Cuicatlán, began with an *apantle* (sluiceway) built in 1846, for a sugar cane trapiche of a man named Rojas.

These monoculture sugar haciendas were populated by Blacks imported to work on them. Esteva tells us that in 1913, in the Cañada,

> hay además algunas familias pertenecientes á la raza etiópica que fueron importados como esclavos durante el Gobierno virreynal, las que se destinaron al trabajo de siembra y recolección de la caña en las fincas cálidas. El Gobierno de nuestro Estado las redimió en 1825, pagando á sus dueños cierta cantidad por cada uno de los individuos pertenecientes á ellos. Hoy existen libres en varios lugares y se han mezclado con los nativos, dando origen á la variedad de *Salto-atrás*. [Esteva 1913:97–98]

The original towns of the Cañada continued to exist and continued to be based on the production of staples and fruit. Cuicatlán maintained a pattern of diversified production with a specialization in fruit. Belmar says that San Pedro produced chicozapotes commercially and that Cuicatlán produced maize, higuerilla, chile (especially chilpete, or chile gordo), and was famous for its ciruelas (Belmar 1901). Cuicatlán also grew some sugar cane although Belmar also mentions Guendalain and Tecomaxtlahuaca as sugar cane producers. A little later, Esteva Cayetano lists fruits grown in the distrito de Cuicatlán, including chicozapote and chile. These communities remained Indian through the nineteenth century, existing side by side with the non-Indian haciendas. The contrast in production paralleled the constrast in structure. There are references to the Indian nature of these towns and to Cuicatec remaining the language in these towns.

When Bandelier passed through the Cañada in 1881, he noted that the Indians (his term) raised two crops of maize a year, since it was *tierra cal-*

iente. He was so impressed with the linguistic and cultural diversity of the area that he wrote: "It is a region which I cannot too earnestly commend to the attention of future students" (Bandelier 1884:266). Unfortunately, few took this advice.

However, these Indian communities were subject to many influences that did not affect the highland Indian communities. First, increasingly, there was a substantial population of Spaniards living in the Cuicatec towns. These were established as merchants dealing in the products both of the lowlands and the highlands. The Cuicatec towns like Cuicatlán were more and more important as administrative centers, because they represented the last point in the ecosystem before transportation became difficult. As a result the Cañada towns played an intermediary role between the highland communities and the Spanish macrosystem. The presence of increasing numbers of people from the Spanish system began to give a two-class structure to the Cañada towns. In short, they were beginning to look more and more like Spanish towns.

At the same time, the highland towns were becoming increasingly polarized from the Spanish pattern. They were relatively (not absolutely) isolated. The relative importance of the highland to the lowland communities began to change, and the lowlands became more important. The lowland population at last began to recover and increased more, proportionately, than the highland population.

This does not mean that the highland communities were untouched by the innovations at the end of the nineteenth century. There were several effects. First, the lowland Spanish interests, in keeping with the increased commerce made possible by the road and later by the railroad, began to reach more directly into the highlands. Some lumbering interests began to establish means to bring lumber down out of the mountains. Late in the nineteenth century, fincas for coffee were established in some of the highland areas. These would have had an impact on the highland communities comparable to the sugar cane haciendas on the lowlands.

There was a nineteenth-century change of the market that acted to impoverish the highland villages. This was the importation of industrially manufactured cotton products and the invention of aniline dyes. These effectively wiped out the centuries-old highland specialization in cotton weaving and cochineal production. While the weaving has survived among other Indian groups in Oaxaca, it has largely disappeared among the Cuicatec. Frederick Starr, who passed through the region shortly before 1900, said of the Cuicatec that, "The Cuicatecs at Pápalo, where we examined them, are disagreeable and uninteresting Indians. They do not present a definite physical uniformity. . . . Few women wear notable dress" (Starr 1900:68).

So the nineteenth century was a resurgence of the lowlands, while the highlands took a back seat. The highlands remained Indian, isolated but somewhat more impoverished. The highland population rose, but the lowland population rose more quickly. The lowland resurgence in part can be seen as related to this population increase. The new return to high density meant that once again the intensive agriculture was profitable, and there was a demand for fruit production. In addition, the new improvements in transportation made the market available for this production larger than ever.

But as we have seen, in this resurgence of the lowlands, a new pattern was emerging. To the extent that the lowlands resurged, they were forced more and more into close relations with the national pattern. This happened in the form of direct establishment of Spanish communities, or haciendas, and in increasing molding of the existing Indian communities in the lowlands towards the Spanish pattern.

The Cuicatec Ecosystem in the Twentieth Century

In the twentieth century the processes that have operated over 400 years of post-Hispanic history have reached their logical end. The differential forces for schismogenesis operating on the Cuicatec ecosystem have resulted in the lowlands becoming entirely Spanish, while the highlands have remained Indian. We can flesh out the outline of the ecosystem that we have been describing with more details from the evidence we have from published material and from observations in the field.

The Cañada towns entered the twentieth century with two forms of pressure on them to change from the symmetrical, Indian structure to the complementary, Spanish structure. The first was the growing competition for the land by sugar haciendas, which were worked by populations of non-Indians. The original Cañada towns were subject to a number of other forces. First there was an increasing number of Spanish administrators, priests, and merchants, who became established in these towns. Commercial production of fruits and vegetables increasingly made these Cañada towns appealing to Spanish landholders. In part, this change in production was related to the increasing population, which again placed demands on the Cañada irrigation system to produce surplus food.

Up to the time of the Mexican Revolution, the competition of the hacienda settlements with the older, mixed-agriculture Cañada towns was lopsided in favor of the haciendas. However, the Mexican Revolution overthrew Porfirio Díaz and changed the extremely "favorable climate" for foreign landholding in Mexico. The Mexican Revolution shows up as a kink in the population curve of the Cuicatec ecosystem.

Another factor may also have helped to stop the increase of the haciendas. That was the changing of the other Cuicatec Cañada towns to Spanish form. The increasing population of all of Mexico, as well as of the Cuicatec area, the improved transportation through the Cañada area (the narrow-gauge railroad through the Cañada built in 1892 was widened and improved in the 1950s), and the increasingly important number of Spanish merchants and administrators living in the Cañada towns have resulted in these towns becoming completely hierarchical, complementary, and national in orientation. Cuicatec is not spoken in the Cañada towns except by recent immigrants from the mountains and some storekeepers selling to Indians shopping in the Cañada towns.

The Cuicatec ecosystem as it exists today illustrates the same processes that we have followed through its history. Today the difference between the lowlands and the highlands has redeveloped as it existed at the time of the arrival of the Spaniards. The differences in economies are tied in with the differences in cultural patterns. The irrigated, commercially-producing Cañada is Spanish, while the largely subsistence-oriented highland systems support an Indian population. This difference is aggravated by the presence in the Cañada of a good transportation link with the rest of Mexico, while the highlands remained relatively inaccessible. The Cañada towns are relatively prosperous, because of their valuable fruit crops, while the centuries-old activities of the highlands, the production of woven cotton and cochineal have been replaced by German dyes and cheap, factory-woven cloth.

The lowland communities grow cash crops, such as tomatoes, chiles, and fruits, especially the mango, entirely with irrigation. The highland people, on the other hand, are largely subsistence agriculturalists, although some special cash crops are grown, especially nuts and apricots. The lowlands, especially Cuicatlán, are the trade centers, where the entrepreneurs live, because of the railroad and the road to Oaxaca. These entrepreneurs have established client relationships with the highland communities. Products to and from the highlands are filtered through them (Hunt 1972).

The Indian communities still remain quite strongly entrenched in their form of life. As we have seen, the forces that worked for the change to the Spanish, complementary structure were concentrated on the lowland, Cañada villages. While the market for highland-produced cotton products disappeared in the last century, Weitlaner (1961, 1969) reports the survival of weaving as a local industry, particularly in Teutila. In Santa María a distinctive wool sarape is still made. Many other ethnic groups in Oaxaca have continued making distinctive garments. However, the Indians of many of the Cuicatec towns, especially the most important ones of the Pápalos and Tepeucila, wore simple white *calzones* rather than any distinctive costume. The women use store-bought *rebozos* (shawls). At present the Cuicatec must be among the least folkloristically interesting of all Mexico's Indian groups.

Weitlaner and Holland (1960) describe the survival of pre-Hispanic curing customs, related to survival of the pre-Hispanic pantheon in the highland Cuicatec towns. Weitlaner (1969) also describes elaborate rituals and paraphernalia for the assumption of office in the same towns. These must have descended from the ceremonies made for the Spanish administrators like those described by Ajofrín (1959) in the eighteenth century. Clearly the conservative Indian pattern is still strong in the highland villages. These highland Indian communities are based on subsistence-oriented agriculture based almost entirely on rainfall. In Concepción Pápalo, there is some very small-scale irrigation from small springs and some rough terracing. In Santa María there are some *bancales* (agricultural terraces) with maguey plants. The irrigation, where present, may be used for seedbeds or for starting crops so that they will be able to mature before frost threatens them.

Corn is, as everywhere, the basis for the diet and the agriculture. The highland Cuicatec towns have many kinds of corn, each with different virtues such as differential tolerance of cold and different lengths of growing seasons. These are sown according to altitude. On a trip collecting corn with Ephraim Hernández X., we collected five different kinds of corn from informants in Concepción Pápalo and three from Santa María Pápalo. Most of these were local varieties, although one of the varieties in Concepción Pápalo was a recently introduced hybrid. In Concepción Pápalo, three varieties of corn were said to be "urbano," that is, planted in the fields in and around the town. One was planted in *clima frío* and one was planted in the lower fields. Interestingly, the one that was said to be planted in the highest fields had the longest growing season, from April to December, while the one that was planted in the lower fields of the towns (at 1200–1400 m) had a very short growing season, from June, July, or even August to December, since the lowest fields are in the rain shadow.

In Santa María, on the east side of the sierra, the two lowland corns which are planted below the town (the town is at 2040 m above sea leavel) have a growing season from April to December, which is as long as the longest growing season for corn in Pápalo. Other corn, planted around the town and higher, has an even longer growing season. It is planted in March and harvested in December. Thus it would seem that the critical variable is the amount of rain available, rather than frost.

Informants in Concepción Pápalo reported two kinds of beans, which were grown between the rows in the milpas. Weitlaner reports that potatoes are grown in two of the highland towns and nuts are grown in Santa María. Santa María produced pulque from large fields of pulque maguey on the slopes above the town. Tepejilotes are also gathered in the highlands.

Some coffee is also grown in the highlands. Apparently the earlier coffee fincas were closely related to the various Cuicatec highland towns as a source of labor and as the fincas failed, the two towns took over the land and kept up small-scale production.

Coffee nuts are sold by the highland towns to the national market. This is done through the merchants centered in Cuicatlán. At the time of the harvests of these products, these merchants send agents up to make a circuit of the highland towns and purchase the crops. They are brought down on pack animals or on trucks through the lumber roads that the Papelería Tuxtepec has constructed through the Cuicatec forests.

The other important activity going on up in the highland villages is logging. The Papelería Tuxtepec has, at considerable expense, constructed an all-weather road from the railroad at Cuicatlán up to Concepción Pápalo. From here temporary logging roads extend up and down through the forests in many directions. The lumber is harvested according to modern tree-farming techniques, including tree nurseries installed in the mountains. The individual Indian towns negotiate the rental of their forest land at regular intervals. The Indians of the highland towns also often work for the paper company. The cut logs are brought down by trucks that belong to entrepreneurs who work out of Cuicatlán. At Cuicatlán the logs are shipped on the railroad.

At the same time, in Reyes Pápalo the older type of logging goes on. The Indians cut vigas (roof beams) and broad axed-out tablas (planks about 6 feet by 20 inches and about 1 inch thick). The Indians then make the trip on foot down the mountain to Cuicatlán with the vigas, tablas, and net-loads of charcoal. They tow the vigas by a string through a hole cut on one end of the beam and carry the tablas and the charcoal with tumplines (Fig. 3).

The lowland towns are entirely irrigated. While there are some small producers who grow corn, much of the land is in the hands of a few wealthy landholders. The important crops are fruits and vegetables planted in the fields and permanent standing fruit trees. The fruit trees can be grown without terraces or with individual half-moon terraces on the steep slopes around the edges of the relatively flat Cañada bottom. Particularly important trees are mangos, chicozapotes, and ciruelas. Trees are also grown in rows between the fields. Sometimes whole fields are planted in fruit trees. San Pedro Chicozapotes is named for the trees that obscure the town from sight, although today there are more mangos than chicozapotes. This is probably relatively recent. Turn-of-the-century sources hardly mention the mango while talking of the

Figure 3. Indians from Reyes Pápalo descending to Cuicatlán with *vigas*, *tablas*, and nets of charcoal.

important fruit production of the Cañada. Today the mango dominates the local economy. In May and June, when the mangos are harvested, it is nearly impossible to hire laborers.

While fruit and vegetables are grown in the fields on the flat Cañada floor, and the fruit trees are grown around the houses, on the slopes below the canals, and between fields, at least in Cuicatlán, the sugar cane was generally grown on the present river floodplain, below the main part of the Cañada floor. The soil the sugar cane is grown in is less loamy and more sandy than that of the Cañada floor. The canals irrigating the sugar cane on the floodplain draw water from the Río Grande. These are possible in part because of the importation of digging machinery, but also in part because these canals do not attempt to raise the water above the first terrace to the Cañada floor. These fields may disappear when the river shifts in flood, as it did in the year I was there.

In contrast, in the Cañada towns that were haciendas, or under heavy influence from the haciendas, almost all the land is in sugar cane. In Valerio Trujano (the former Guendalain) and in the lands of San Pedro Chicozapotes that were previously the hacienda of Tecomaxtlahuaca, sugar cane is still almost the sole crop. Significantly, these two towns now have a high Black population, the descendants of Blacks imported for sugar cane production. The refinery of La Iberia, near Cuicatlán, which once not only refined sugar but produced industrial alcohol and a local rum, has been shut down now, perhaps because of the improvements in the railroads in the 1950s. Now the cane is taken to Orizaba in the form of raw stalks for processing.

There are three kinds of irrigation in use in the Cuicatec ecosystem. Irrigation, where present in the highland communities, is usually in the form of small trickles leading to individual plots from small springs. This irrigation is usually found in the fields in or near the towns, where slightly more intensive agriculture is done. This water may help get crops started early enough that they may mature before frost sets in.

Intermediate between this and the irrigation on the Cañada floor is a small irrigation system above Dominguillo in the town of Santa Cruz. This small system uses the copious flow from a spring in a cave just above the town. This water flows down through canals by the side of the mineralized remnants of earlier canals, like those on Hierve el Agua (Neely 1967; Flannery and Marcus 1976; Hopkins 1983). Installed at the upper edge of the fields is a series of tanks. These store water for when the flow of the spring diminishes in the dry season. This town is currently inhabited by some 30 people. Immediately above it is a large archaeological site.

Finally, in the Cañada, fields are entirely dependent on irrigation like that practiced in the town of Cuicatlán, based on water from a Río Chiquito or from the Río Grande. The older and more traditional system appears to be the Río Chiquito system.

The Río Chiquito system takes water from the small tributary river that descends between Concepción Pápalo and Reyes Pápalo. On both sides, the water is taken off high up on the river, where its canyon is quite closed in. The south *toma de agua* (water intake) is the lower of the two. The river runs closer to a rock wall on that side, which would make taking the water higher up difficult. The canal from this *toma* waters the town and the land to the point where water from the Río Chiquito de San Pedro meets it. The *toma de agua* on the northern, or La Sabana, canal is considerably further up the Río Chiquito.

The water from the Río Chiquito is diverted into the mouths of these canals by means of brush dams. These dams are constructed with teepee-like upright structures which support other brush and rocks to form the dam (Fig. 4). These dams are reconstructed every year and gradually fill with sediment. Every year in the rainy season the dam is washed out. At this time the water level is high enough to enter the mouth of the canals without a dam. When the water level falls, a new diversion dam is constructed. At present the mouths of the canals are reinforced with a concrete structure at the upstream end of the mouth. These are relatively recent. The one on the *toma* on the south, town-side canal dates to 1965, according to an inscription in the concrete. Above this *toma*, in the miniature and quite constricted floodplain of the Río Chiquito, is one small canal and rows of lemon and banana trees; these must survive the annual floods.

The southern, Cuicatlán-side town canal is simply a ditch that carries the water around at a constant drop to the fields below the canal, although there are some stone-lined channels within the town proper that guide this water. The canal on the northern side in two places passes over well-con-

Figure 4. Brush dam construction for *tomas de agua*. *Above*, Río Chiquito, Cuicatlán; *below*, Río Grande, Cuicatlán.

structed brick aqueducts that seem to date from an earlier, but post-Hispanic, period. This canal is called the Apantle Nuevo, although no one could remember when it was constructed. At one point on this canal there is also a cement overpass to carry the runoff from a gulley over the canal. The date 1946 appears in the cement.

Most of these canals are just earth ditches which run along the contours with a constant drop sufficient to keep the water running briskly. I was told that they are dug by a sort of trial-and-error technique. A section is dug, water is let into it, and then it is raised or lowered according to how the water runs through it. Water is taken off these canals through places where the downhill side of the canal has been broken through. These holes are plugged with stones when water is not being taken out. There are also major branches of the canal that descend to lower fields. These are usually equipped with a more formal concrete sluice, often with a padlocked gate. These are quite simple constructions of concrete with a blacksmithed metal gate.

Where the water is let out directly from the main canal on the talus slopes planted in trees, it is usually led down steep little canals to fruit trees planted on the slopes, going back and forth through a series of switchbacks across the slope from tree to tree. The trees have a little bank around them that allows a pool to accumulate around them. Often the trees are in individual half-moon terraces built up of rocks.

Further down in the system, when the water is let into fields, individual canals are tended in the fields by men with hoes and shovels who carefully see that the water flows evenly and gets to the whole field. There are terraces in some of the steeper fields to even out the irrigation, and even some wet rice paddies, but mostly the water is led through the fields, gradually descending through the contour furrows. Some seedbeds are used to start plants before they are planted in the fields.

The systems watered by the Río Grande are distinct from those watered by the Ríos Chiquitos. They occupy the present floodplain of the Río Grande, while the fields serviced by the Río Chiquito canals are up on the river bench (see Fig. 5). The river bench edge in the irrigation area of Cuicatlán is marked by the railroad track. There is a distinct 5 m drop from the bench to the present river floodplain.

The canals from the Río Grande only irrigate land on the floodplain. The Cuicatlán canal is dug by a back hoe, and the dam is bigger, since it reaches into the Río Grande. It is of the same brush construction, however, as the *tomas de agua* on the Río Chiquito (Fig. 4). As the Hunts have pointed out, this land is all in the hands of the large landholders and is almost all planted in sugar cane (Hunt and Hunt 1973).

The sharp contrast between the motley pattern of fields above the railroad and the river bench, and the uniform fields of sugar cane on the lower floodplain can be seen on aerial photos of the Cañada. The lower sugar cane fields are in danger in the rainy season floods when the river changes its course, as it did in 1969. At this time one landholder lost half a hectare to the river. Most of the floodplain fields were inundated and their crops lost. At this time the Río Grande canal was completely wiped out, and trees up to 2 m in diameter were carried off. The last previous flood of this magnitude was in the 1940s. Because of this, the occasional attempt to plant fruit trees on the lower floodplain is probably a risky proposition.

In the Cañada towns where all land is not devoted to sugar cane, there are three classes of land with their respective crops (Fig. 5). On the steep slopes below the canals, fruit trees are planted, sometimes on individual half-moon terraces. On the less steep slopes, sometimes on sloping terraces and on the flat of the Cañada bottom, crops are sown, with fruit trees along the canals and field boundaries. Depending on the markets, sometimes entire fields are planted in trees down on the flat. At the time I was there, people were planting fields of mango at the expense of other crops. Until the trees are big, the areas in between the trees can be cultivated.

Finally, in the floodplain of the Río Grande, the sandy, gravely soil is devoted to sugar cane. Sugar cane does not seem to be grown on the flat ground very much except in ex-hacienda areas. While Eva Hunt attributes the fact that this sugar cane land is owned by the wealthy land owners to their monopolizing the best soil, I suggest that she was closer to the truth when she recognized the need of greater capital to exploit the Río Grande. This water source is more than adequate to supply enough water for all the sugar cane, which demands a relatively large amount of water, that can be grown on the floodplain. As long as no effort is made to raise the canal up the 5 m river bench, no considerable engineering feat is required, beyond the reinvest-

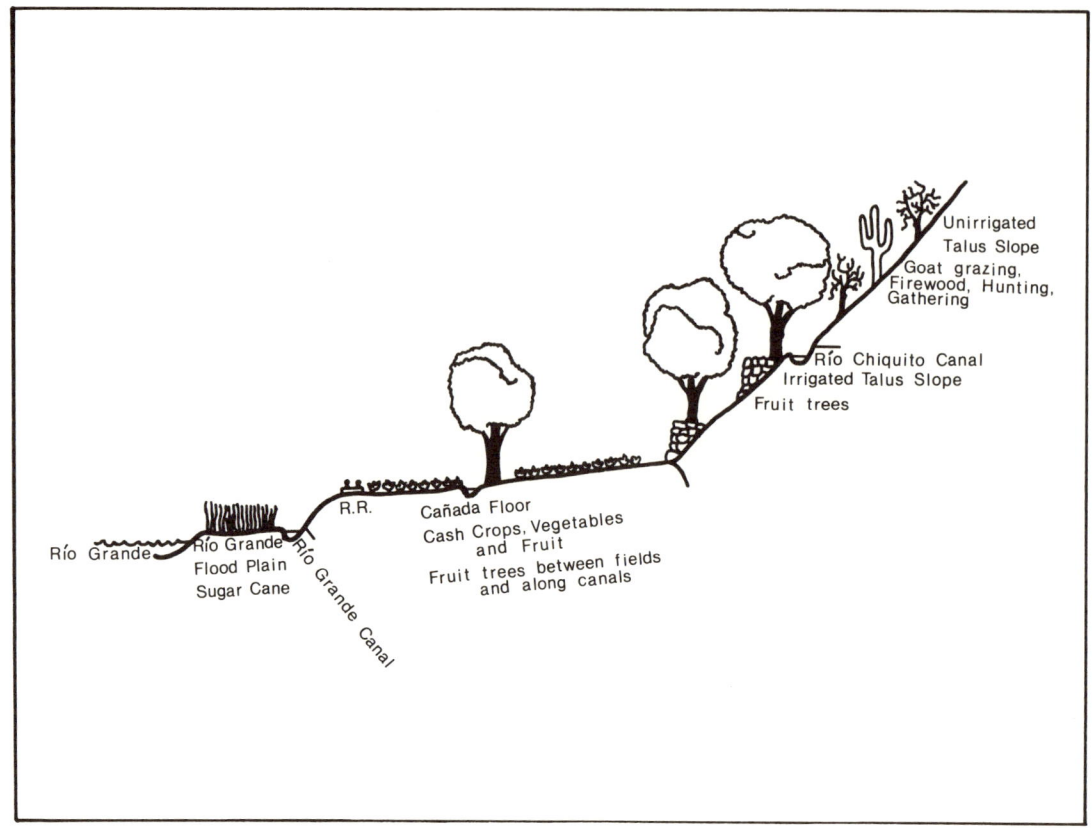

Figure 5. Land use in Cuicatlán.

ment of capital, labor, and rope to rebuild the dam every year, and occasional redigging of the canal after exceptional floods.

A smaller Cañada irrigation system was visited at Atlatlauca. Here one did not have the feeling that the same kind of population pressure was working on the system. There seemed to be enough water for the number of people using the system. The canals were not so carefully graded to keep the canal as high as possible. They were smaller, and where it was inconvenient to keep the canal up on the slope, it was simply allowed to fall. One element not seen in the Cuicatlán system was the use of *canoas*, hollowed-out logs, to carry the water across streams (Fig. 6). This system may approximate the way the Cuicatlán system looked in the eighteenth century when its population was lower.

Eva Hunt (1972) had described the kind of political structure by which the canals are administered. There are three different systems for three separate canal systems. The town canals that are taken off the Río Chiquito are administered by a Junta de Aguas, which is under the national department of Recursos Hidráulicos. The canal taken off the Río Grande is administered by the rich landowners who built it and who are its chief users. And the canal that irrigated the ejido is administered through the ejido organization. The rich elite of the town is able to dominate two out of three of these because they in fact put their clients in positions on the Junta de Aguas. The municipal authorities are not thought to be very important people because they are recognized to be puppets of the elite.

Each canal system has its own governing body, which is smaller than the local political unit. The means of establishing clientship relationships within the town is only partly tied to the irrigation richness of the elite families. In part, it is also due to their commercial activities, which extend out into a region-wide system, including supplying and trading for products from the highland towns, as well as through the lowland area. Thus, at least two of the elite families run stores and regularly send trucks and pack animals up into the mountains to

Figure 6. *Canoas*—hollow treetrunks used to carry canals over barrancas at Atlatlauca.

sell goods and buy up highland specialty crops. They also extend credit to people within the town, who then are dependent during the bad times on the storekeepers-landowners. In addition there are a number of people who depend on salary labor on the land of the elite.

Analysis: Processes in the History of the Cuicatec Ecosystem

Following Hans Jenny (1958, 1961) we have defined an ecosystem around the Cuicatec region. We defined the time just before the Spanish Conquest as the initial state of the ecosystem and followed the ecosystem to the present. We have fixed the arbitrary boundaries of our ecosystem around the towns that were Cuicatec at the time of the Spanish Conquest.

The individual towns are the units within this system. Initially there were two kinds of units, highland and lowland. With the arrival of the Spaniards, a new kind of unit developed in the Cañada, monoculture sugar cane haciendas, which existed alongside the older Cañada towns.

I have analyzed the structure of the ecosystem in terms of Bateson's article (1972) on culture contact and schismogenesis. First, the individual Cuicatec towns had a complementary structure within them, with a noble and a commoner class, related in a hierarchical fashion. Relations between the towns within the ecosystem were either symmetrical (as in the case of the intermarriage of the nobles from village to village), or complementary in a way that reinforced equilibrium, such as the highland-lowland exchanges. The same was true of the relations with other towns outside the ecosystem (to the extent that the boundaries of the ecosystem were permeable).

There was one important set of relations of the ecosystem to its macrosystem that was not symmetrical. This is the relation of the ecosystem to the Aztec system. This relationship was a complementary one, which paralleled the relationships of the nobles to the commoners within the individual towns (indeed, the structure of the towns may have been in imitation of the larger structure). The Cuicatec were tied through Coixtlahuaca to the Aztec "empire" and had complementary, hierarchical relations with them, with the Cuicatec on the bottom and the Aztec on the top.

Understanding this structure in which the ecosystem was set is essential for understanding the nature of the ecosystem and its development. As Wolf says,

> nations or "systems of the higher level do not consist merely of more numerous and diversified parts," and . . . it is therefore "methodologically incorrect to treat each part as though it was an independent whole in itself" (Steward, 1950:107). Communities are "modified and acquire new characteristics because of their functional dependence upon a new and larger system" (ibid:111). [Wolf 1956:1065]

The Cuicatec ecosystem is one of a number at the bottom of a hierarchy. In the initial state, this consisted of a connection through Coixtlahuaca to the Aztec capital. Decisions could be made in the macrosystem that arbitrarily determined what happened in and to the ecosystem. Secondly, for the time we are considering this ecosystem, innovations always originated outside in the macrosystem. From the outside, these penetrated the ecosystem, where their effect then was determined by the structure of the microsystem. Thirdly, one of the constant factors of the macrosystem, since it was hierarchical, was that it demanded tribute, taxes, or some kind of constant outflow from the ecosystem. This demand has to be taken into account to understand some of the characteristics of the ecosystem. In order to administer and enforce this drain, there was also an input from the outer system to the ecosystem in the form of administrators, priests, etc., who had their effects on the workings of the ecosystem.

The interchange of highland products and lowland surpluses and the necessity of irrigation in the Cañada villages were clearly closely related to and constrained by the structure of the natural system. Some of the other aspects of the adaptations of the highland and lowland towns seem to be related to the complementary interchanges themselves, but not obviously to any environmental constraint. The highland specialty of weaving cotton that could only be obtained from lowland villages and then trading the finished product back to the lowlands is an example.

The interaction of the natural system and cultural system in the ecosystem explains the form of the ecosystem as it existed at the time of the Spanish arrival. Within these two subsystems we see that (1) relief, (2) the complementary interchange of products, and (3) the demands of the Cuicatec ecosystem are the results of the complementary relationship, with the Aztec part of the macrosystem playing particularly important roles. These factors continued to be important through the entire history of the Cuicatec ecosystem.

The Spanish Conquest resulted in important changes in the Cuicatec ecosystem. The depopulation that followed the Conquest had a strong impact on the natural system of the Cuicatec ecosystem. The Spaniards also caused important changes in the structure of the cultural system and introduced several new elements to this system.

Much of the natural system remained constant from pre-Hispanic times through to the present. The climatic patterns, the relief, the various natural plants and animals available in the area, all remained the same. However, the introduction of epidemic diseases to which the native population had no resistance had an extremely important effect. This depopulation and subsequent recovery had a profound impact on the developmental processes of the Cuicatec ecosystem from the time of the arrival of the Spaniards to the present. The depopulation and the later recovery has been a factor for the entire post-Hispanic history of Mesoamerica. It seems that only in the present century is the population returning to pre-Hispanic levels. The pattern of this depopulation is summarized in Figure 2 and in Table 3.

While I have not tried to make a numerical estimate of the population of the Cuicatec ecosystem at the time of the Spanish contact, it can be stated confidently that the population was quite high. We know that thereafter the population dropped considerably and then rose. This suggests a corollary to the Ester Boserup hypothesis (1965:63) that intensive agriculture is the result of increase in population. Therefore a decrease of population should then lead to a reduction in the intensity of agriculture. This should presumably have affected the Cañada, which began with more intensive agriculture and suffered a greater drop in population, more than it would have affected the highlands. The small-scale highland irrigation might also be expected to disappear as the population drops.

At the time of contact with the Spaniards, the population was undoubtedly at least as large as the present population. As a result of both high population and the demands of the Aztec tribute, the Cañada towns were quite important. With high population everywhere the intensive food production and guaranteed surpluses gave the Cañada towns considerable leverage on the other towns. We know that Pápalo counted on Cuicatlán for food when her crops failed, and that several other towns either rented lands in the lowlands or worked as laborers in the Cañada in order to make their tribute. We know that Almoloyas had to go to war to guarantee their supplies from the lowland Cañada towns under their sway. The high population made this intensive food production in the Cañada towns necessary and economical. There were enough people within reach, even with a primitive means of transportation, that it was economical for the Cañada towns to sell their produce throughout the region. This, in turn, gave them considerable importance. The highland slash-and-burn or fallowing system did not have the same potential for intensification.

The depopulation was selective in its influence on the highlands and the lowlands. While the figures for Mesoamerica as a whole are more extreme, our population figures show that the Cañada towns lost a higher percentage of their population than the highland communities. The highland communities also seem to have begun recovery earlier than the lowland communities.

This effect was twofold. For one thing, the simple discrepancy of the relative losses had to have affected the relative importance of the towns. Secondly, the general sparseness of the population had to have made it more difficult for the Cañada towns to find a market for their fruits and vegetables. The reduced population pressure on the highlands should have reduced their need to depend on lowland towns for their subsistence and tribute. This, in turn, must have reduced the political importance of the lowland towns.

In fact, we do find that for the period of low population the lowland towns are reduced to subsistence agriculture, with little or no exportation of fruits. The highlands at this time were still relatively important because of their trade in cotton, grana, and silk. These products did not depend on a nearby local market and were valuable enough to be exported a good distance.

However, as the population had gradually (and recently, rapidly) increased, this process has been reversed. The Cañada towns are now exporting cash crop fruits and vegetables. They even import some corn, so they can grow other more intensive and valuable crops. Now, for various reasons, the highland towns are expanding their fields to the lower limits of where agriculture is possible, and many of the highland residents are descending to work as paid labor in the Cañada towns. While they state that they do this to avoid the corporate *tequio* labors in the highlands, surely lack of land and a good way to make a living must be factors.

This corollary of the Boserup hypothesis may have validity in the Cañada irrigation systems. This raises interesting questions about the usefulness of the hypothesis for explaining the origin of these systems. We will examine these in the next chapter.

The Spanish Conquest resulted in important changes in the structure of the cultural system of the Cuicatec. It did not result in the total destruction of the pre-Hispanic system. Instead, an opposition was set up between two kinds of cultural patterns, and a process of schismogenesis from one to the other began. Since this process has not come to an end after 400 years, we cannot assume that the end result will be elimination of one of the two patterns. Rather, we must also look for the forces which tended to maintain the opposition.

This process of schismogenesis began by raising the level on which complementary structure existed one step. The complementary relationships within the Indian towns that existed at the time of the Spanish Conquest disappeared by the end of the sixteenth century, and from then on Indian towns were characterized by a symmetrical class structure. Other towns with a considerable or even small Spanish population showed the complementary class structure that had characterized the pre-Hispanic Cuicatec towns. At the same time the Spaniards replaced the Aztec in the complementary relationships between the macrosystem and the Cuicatec ecosystem. The Spanish pattern was characterized by complementary relationships, at all levels, while the Indian pattern was characterized by symmetrical structure on the local level and with the inferior position in complementary relationships with any Spaniard. The Indian pattern was always local, and hence symmetrical, while the Spaniards always were related to the macrosystem, and hence were part of a complementary system.

This more complex structure evolved during the sixteenth century. By 1600 the Cuicatec ecosystem was solidly Indian, in the new sense, with symmetrical organization, under a Spanish (complementary) macrosystem. Through the rest of the history we will see the Spanish pattern interacting with the Indian at the latter's expense.

The Spaniards introduced a number of new elements to the Cuicatec ecosystem. These elements were adopted or rejected in complex ways according to whether they were more compatible with either the Spanish or the Indian systems, as well as with the different possibilities of the natural system. Elements often were specifically associated with one or the other system and as such were resisted or embraced with vigor by the people in the Cuicatec ecosystem.

Some elements were tied to the Spanish cultural system. Their acceptance tended to cause a change in the equilibrium towards the Spanish side in towns where they were adopted. For example, changes in transportation meant the greater intrusion of the macrosystem, which, as we have seen, was complementary and Spanish, into the Cañada. Again, any introduction such as sugar cane that demanded a complementary structure for its exploitation tended to lead to the change to the Spanish pattern. And finally, later innovations, coming from the outer system as they invariably did, represented a dependence on the Spanish, complementary system, and tended in the long run to be forces towards the Spanish end of the opposition.

This structural opposition was established in the sixteenth century. First, the Spanish officials were established who took over all the roles of the Indian aristocrats, in religion and in the government of the towns. These were the first priests, monks, and encomenderos. This resulted in the erosion of the power of the Indian aristocrats and their submergence into the mass of Indians. However, the Spanish direct influence waned early in the Cuicatec region. As a result we find that by the end of the sixteenth century the Indian villages are symmetrical and largely untouched by direct Spanish contact, save by occasional visits by priests and tribute collectors.

While this social change was going on, aided by the massive depopulation that decimated the Indian aristocrats and their subjects, making the land marginal for supporting Spanish administrators, a number of elements were introduced by the Spaniards. These included new plants and agricultural techniques; changes in technology, especially transportation; and the effect of technological changes on the external markets to which the Cuicatec ecosystem sent its products.

Where the Spanish innovations were worked into the pre-Hispanic Indian system, they constituted additions. Thus, goat herding seemed to blend in easily with farming, as did chickens and pigs. Where flat land permitted the use of the plow, it was used, although this seems to have been accepted to a greater degree in the fields of the lowland Cañada towns than in the steep mountain fields. Similarly, many crops could be added to existing crops. At first there were more crops accepted in this way in the highlands, simply because the Spanish crops were temperate. Later many tropical fruits suitable to the lowland Cañada towns were introduced.

Another crop, sugar cane, which was planted for its commercial value, was part of a complementary and ultimately Spanish system, unlike the fruit and vegetable or subsistence production characteristic of the Cuicatec Indian ecosystem. Sugar cane could not be grown in the highland communities, but had a considerable impact on the lowland Cañada towns.

Land in the Cañada tended to be either in one system or the other. Thus the sugar cane land in Cuicatlán today, which was always a mixed producer, occurs below the river terrace, where it can be irrigated by the Río Grande. On the other hand, in areas of old haciendas, land is almost entirely devoted to production of sugar cane to the exclusion of any other crop. Sugar cane is tied in with wage labor; ownership of land by an absentee owner; and, often, a Black population descended from slaves who worked the sugar cane. A good example of this, across the Río Grande from Cuicatlán, is the town of Valerio Trujano, the former hacienda of Guendalain. Coffee may have started to have a similar effect around the end of the nineteenth century in the highland Cuicatec communities.

Transportation is another factor that was always important in the Cuicatec ecosystem. This factor was partly a function of relief. However, during the history we have considered, there have been major changes in transportation due to technological innovations introduced from the macrosystem, or decisions about investment in the transportation system that passes through the Cuicatec ecosystem. These decisions have been both positive and negative. They have had marked effects on the Cañada part of the ecosystem.

The lowland Cuicatec towns lie along a major north-south route through Mesoamerica. The highland towns are difficult to reach, both from the lowlands and from one highland town to another, due to the mountainous terrain in which they are located. As a result, a part of the importance of the Cuicatec Cañada towns has been a function of their location on this route. Even when their population was at the lowest, the Cañada towns were still nodes on a travel network and the home of the administrators. Through these lowland towns would pass the products from the highlands that were destined for other points in Mesoamerica. This then was a long-term influence in the macrosystem and hence towards the Spanish pattern.

The initial introduction of draft animals would have had a quantitative but not necessarily a qualitative effect on the transportation system. The draft animals may have made longer trips to market more economical. Relays of porters were replaced with relays of horses. As late as 1777, when Thiéry de Menonville passed through, the road was only passable on foot, or by animals, but not for wheeled vehicles.

Juárez improved the roads through the Cañada in 1850–65. If this then made it possible for wheeled traffic to pass through the Cañada, the technological improvement would have an effect on the Cañada towns.

Previous to the improvement the major route south to Oaxaca split at Dominguillo. During the dry season, the best route was up the canyon of the Río de las Vueltas to Atlatlauca, then on up and into the Etla arm of the Valley of Oaxaca at San Juan del Estado. In the wet season, the stretch from Dominguillo to Atlatlauca, where the river filled the canyon from wall to wall, became impassable, and the route ascended at Dominguillo, coming out at Huitzo and entering the Valley of Oaxaca there.

When the permanent road was constructed the latter was the route that was followed. This left Atlatlauca isolated. Since at this time Atlatlauca was barely beginning to recover from its low population, it remained out of the mainstream, much as the highland communities were.

The construction of the railroad in 1892 exaggerated the differential effect of transportation on different towns in the Cuicatec ecosystem. This transportation differential between the highlands and the lowlands at first fostered the tendency of haciendas and fincas to predominate over the older, pre-Hispanic style mixed agricultural patterns. The period of the construction of the railroad represented the height of the predominance of the hacienda. This in part was also a function of a national political atmosphere. In the highlands, it was arrested by the Mexican Revolution. However in the lowlands, the areas where sugar cane got a foothold at this time never really returned to the old pattern.

A third effect of the technology has been the effect of the exterior market on the ecosystem. A much larger market for the fruits, vegetables, and sugar cane opened up as a result of the improvements in transportation. At the same time some innovations have had a negative effect on some of the special products of the highlands. Grana was essentially replaced by aniline dyes. The market for fine handwoven cotton products was destroyed by the market being flooded with cheap English and American factory-woven cloth. Where now one could argue that there is a resurgence of a market for these because of tourist trade, the poor transportation potential for the Cuicatec highlands has weakened the area's ability to contribute or even to recognize the existence of the market. So while modern industrialization has introduced innovations, their effect has been to advance the Cañada while holding back the highlands. On the other hand, the introduction of a road, trucks, and modern scientific logging to the highlands may change this imbalance.

Interaction of Factors

The actual development of the Cuicatec ecosystem from the time of contact to the present has not been the product of the individual effect of these variables, but rather an interaction of these factors with each other and with other factors that were less important or constant over the life of the ecosystem. Having discussed the easily isolated factors, I will now try to described the development of the system with the interaction of all the factors. I have already sketched the system as it stood at the time of contact, so I will start with the impact of the Spanish contact.

The first and most immediate effect of the Conquest was the beginning of the drastic depopulation. Because this effect took place so quickly, it determined the effect of other introduced elements, in some cases delaying their impact, and in others facilitating their adoption. Thus, depopulation probably kept wheat and many other Spanish crops and encomiendas with large livestock ranches from appearing in the Cañada or the Cuicatec highlands, since there was plenty of land nearer the concentrations of Spaniards. The small Cuicatec ecosystems had little to offer to the Spaniards. The low population may have reduced the pressure on land in the lowland communities, in such a way as to facilitate the introduction of goats to Cuicatlán. And at a later period, the low population, again, in the lowlands, and to a certain extent in the highlands, may have made possible the takeover by haciendas of considerable parts of the land for sugar and coffee fincas in the nineteenth century and on into the twentieth century.

The Spaniards also introduced some major changes in the market. They introduced gold currency, which would not have affected the Cuicatec ecosystem, since their one gold mine disappeared early. Feathers, and featherwork, which had been a major highland activity, became unimportant; but, silk was introduced, and the importance of grana and cotton weaving was increased (Hamnett 1971).

As a result, the first new stage of the system was the same time as the population nadir. The lowlands were depopulated proportionately more than the highlands. More importantly, the depopulation affected the basis for the importance of the Cañada towns, that of selling their agricultural products to the surrounding population. In the meantime, the highland communities found their agricultural fields were now adequate and were able to pay their tribute by raising grana and silk and weaving cotton. The result of this is an increase in the importance of the highland towns relative to the lowlands.

At this time all of the towns of the Cuicatec ecosystem were symmetrical Indian towns. The highlands were linked to the complementary Spanish system through trade in cotton and grana, while the lowlands dealt with the Spanish system only by providing relays of horses for travelers through the Cañada. At this point, when population was at its lowest, sugar cane and the haciendas gained their foothold in the lowlands. From this time on, the trend began for the reemergence of the Cañada towns and the decline of the highland towns. At the same time the sugar haciendas first rose and then declined in importance. These can be traced to the dual effects of increasing population and many technological innovations. These signaled the glorification of at least some Cañada towns and the decline of the highlands. They also moved the lowlands further towards the Spanish structure.

Through the nineteenth century the population continued to rise. The irrigation system that had previously shrunk in on itself with the deintensification of agriculture now began to expand, in particular on the haciendas. The haciendas expanded into the land once occupied by lowland fruit and vegetable irrigated system.

At the same time changes in technology resulted in the elimination of a market for cochineal and handwoven cotton. At this time the changes in transportation, culminating in the railroad, gave a further impetus to the growth of the monoculture haciendas and the importance of the Cañada towns near the railroad. Atlatlauca, cut off from all of this, never really became important, but remained a relic of earlier times when the population of the lowland towns was still quite low. It is still partly Indian, while all other Cañada towns are primarily Spanish.

As these trends continued, the hacienda bubble burst, in part because of the Mexican Revolution. As the Mexican population had grown astronomically, the sugar mill in Cuicatlán is closed. The fruit and vegetables are more profitable today than sugar cane.

At the same time the ecosystem is much more part of the national system because of the improvements in transportation. Because of its position on the road and the railroad, Cuicatlán finds that government is a major activity and source of money for the town. Recently, because of the road built by the paper company to the highlands and some new airstrips, schools have been established in the highlands. However, all these influences follow the pattern of entering the highlands through the Cañada towns.

Through the historical sources we have developed a model for the working of the Cuicatec ecosystem. To the extent that we understand the workings of this system, we can extend this model back to archaeological times. Through archaeology we can deal with other aspects of the ecosystem we have observed through history.

The historical data cover the period from just before the Conquest up to the present. In order to analyze the Cuicatec ecosystem from the Conquest back in time, we must use archaeological evidence. This evidence takes us back from the Conquest to the time of the beginning of the ecosystem. Through archaeological evidence we can see what conditions preceded human colonization of the Cuicatec area and what was the effect of the human occupation. Secondly, we can follow the processes of the development of the system from the beginning of the system to the Conquest.

By using the ecological frame of analysis, we can follow the same interaction of factors in the archaeological data that we traced with the historical analysis. We can demonstrate the identity of the processes we find from the archaeological evidence with the same processes as seen from historical sources because the two meet at the time of the Conquest. In the next chapter, we will push our analysis back in time through archaeological evidence, using the model that we developed from the historical sources in the two preceding chapters.

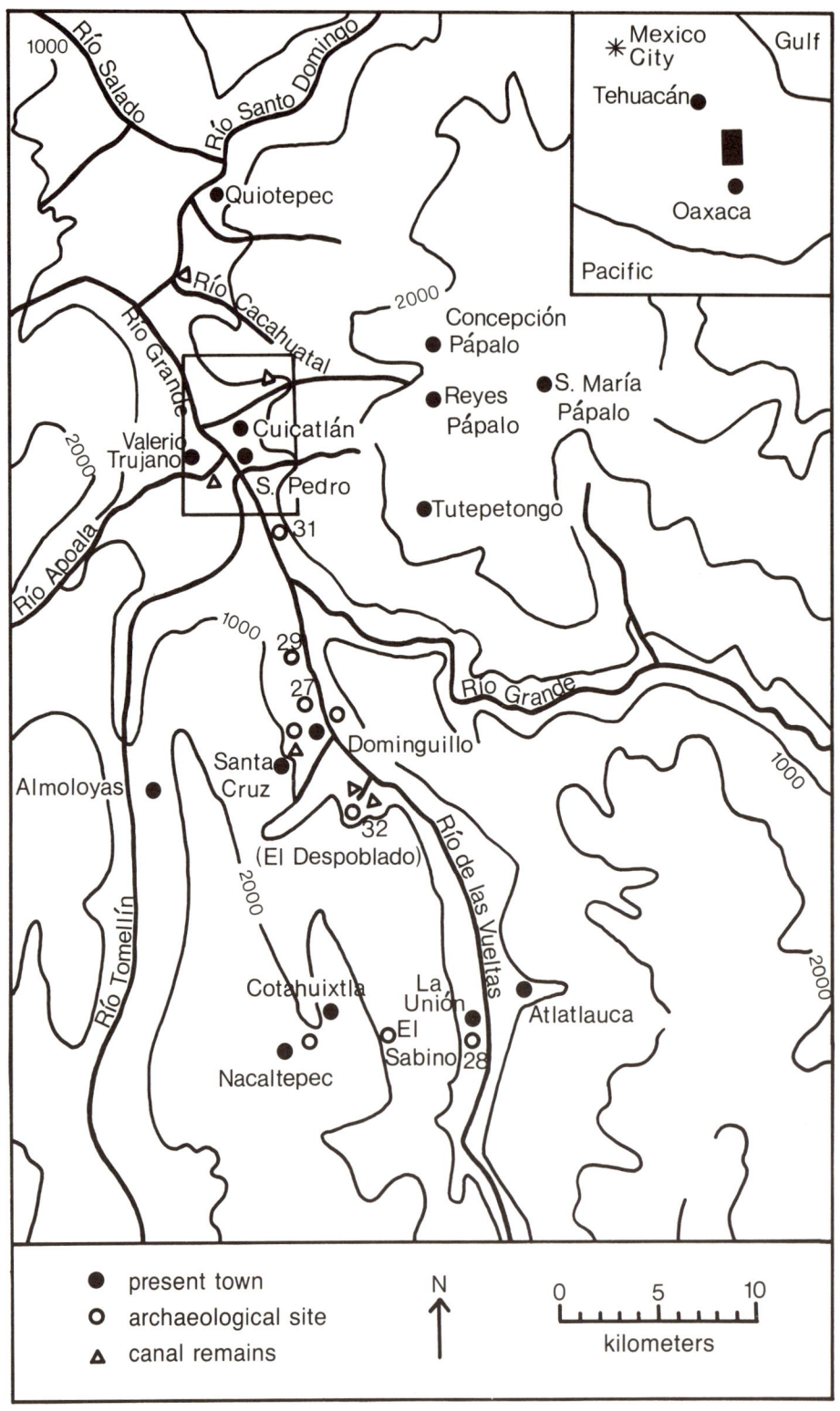

Figure 7. Archaeological sites in the Cuicatec region.

Chapter 5

Archaeological Evidence:
The Cuicatec Ecosystem from its Origin

To the north, south, and west of the Cuicatec region, the Tehuacán Valley, the Mixteca Alta, and the Valley of Oaxaca are known as much for the excellence of the archaeology carried out there as for the importance of these areas in Mesoamerican prehistory. To the south, the Valley of Oaxaca has been the scene for projects by Bernal, Caso, Paddock, Flannery, Marcus, Blanton, Kowalewski, Feinman, and others. Other early work by Bernal and Caso was completed west of the Cuicatec region in the Mixteca Alta. The University of the Americas has also worked there. More recently Ronald Spores (1972, 1983) brought an interdisciplinary project (combining archaeology and ethnohistory) to the Mixteca Alta. To the north, the Tehuacán Valley hosted the Tehuacán Archaeological Botanical Project, headed by Richard S. MacNeish (MacNeish et al. 1972), one of the first large-scale interdisciplinary projects on which later projects such as that of Flannery (Flannery, Marcus, and Kowalewski 1981; Flannery and Marcus 1983) in Oaxaca were patterned. The Palo Blanco Project, directed by Robert D. Drennan (1977, 1978, 1979), continued this research in the Tehuacán Valley.

Until recently, the Cuicatec area had been as neglected archaeologically as it was slighted in many of the historical sources. Pareyón completed one short season's work at Quiotepec (Pareyón 1960). Bernal (1966) published a description of a two or three-day visit to the site of Santo Domingo, in the hills above Dominguillo. Seler (1906) published a brief note on one vessel supposedly found in Cuicatlán, from a private collection. In addition there are unpublished brief surveys in Cuicatlán by Pedro Armillas, Robert and Eva Hunt, and another investigator on a small Wenner-Gren Grant.

From 1968 to 1970 I conducted fieldwork in the Cuicatec Cañada (Fig. 7). This work consisted of an intensive surface survey in the area of Cuicatlán, a ten-day survey of the rest of the Cañada area, a survey of two highland towns, and two small excavations. This work focused on Postclassic remains that dominated the area around Cuicatlán (Fig. 8). This fieldwork resulted in my Ph.D. thesis (Hopkins 1974).

After I left the field several important field projects filled in the picture of the prehistory of the Cañada. Elsa Redmond conducted a survey of the main part of the Cañada, from Dominguillo to Quiotepec (Redmond 1983). This was complemented by excavations in the Cañada by Charles S. Spencer (Spencer 1982; Spencer and Redmond 1979, 1983). In addition Adriana Alaniz of the Centro Regional de Oaxaca, I.N.A.H. conducted excavations at the Rancho Dolores Ortiz (*Boletín* 1975). This work, together with mine, allows us to piece together a good picture of the development of the Cañada irrigation systems.

I will sketch briefly the outlines of the Preclassic and Classic, fitting sites from these periods located by my survey into this sketch. These periods are dealt with much more thoroughly elsewhere (Spencer 1982; Spencer and Redmond 1979, 1983; Redmond 1983). In this chapter I will emphasize the Postclassic remains that correspond to the ethnohistorical model I have contructed above.

Redmond (1983) and Spencer (1982) have divided the ceramic sequence of the Cuicatlán Cañada into four periods: Perdido, Lomas, Trujano, and Iglesia Vieja. The Perdido phase is late Middle Formative, dating from about 650 B.C. to 200 B.C. The Lomas phase is Late and Terminal Formative from 200 B.C. to A.D. 200. The Trujano phase is Classic and possibly extends into the Early Postclassic. It begins around A.D. 200 and lasts until ca. A.D. 1000. The Iglesia Vieja phase is Postclassic and continues from the Trujano phase to

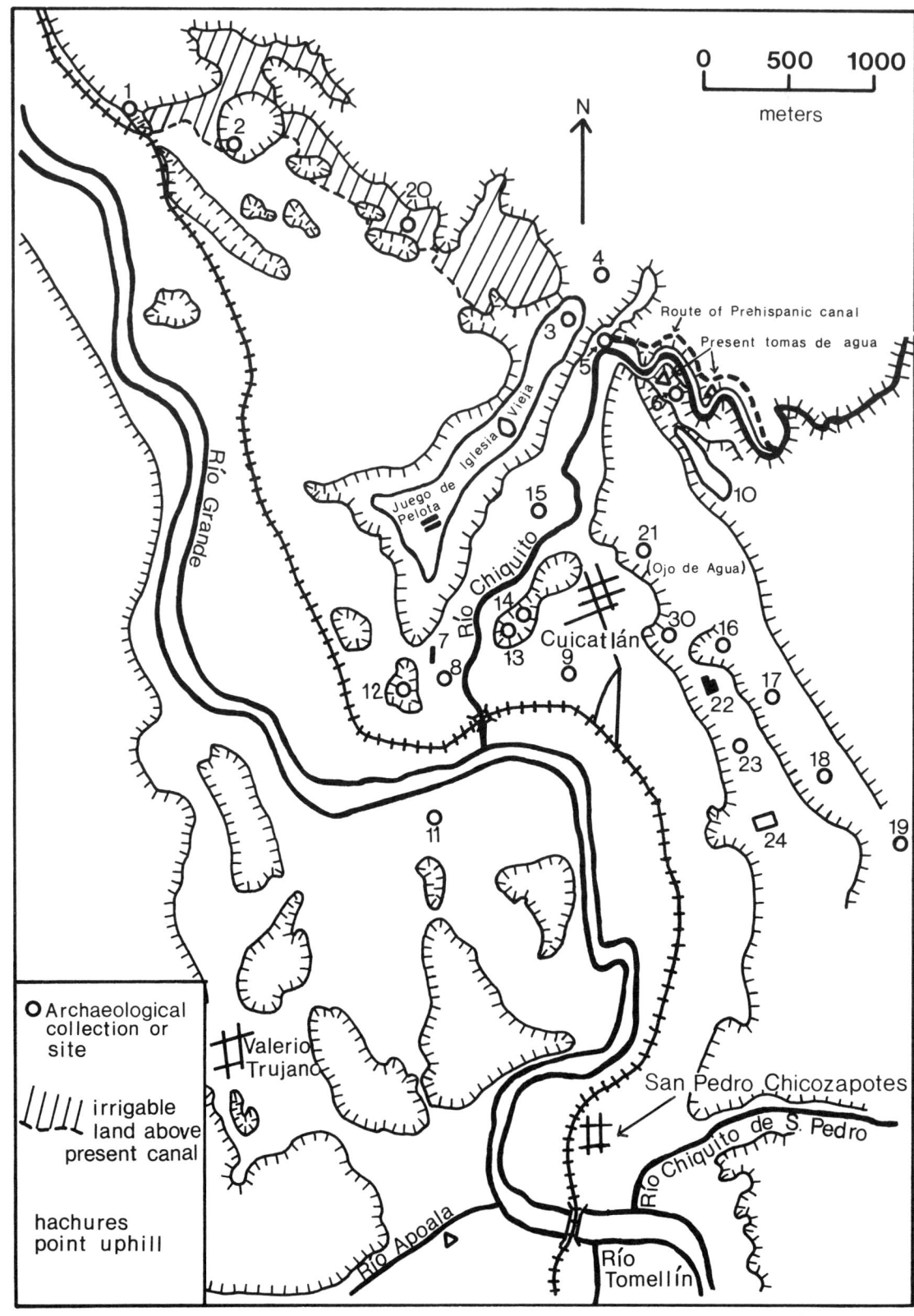

Figure 8. Sites in the vicinity of Cuicatlán.

the time of contact with the Spaniards in the fifteenth and sixteenth centuries.

I will summarize the archaeological evidence for the Cañada, beginning with the earliest remains. I will relate this evidence to the model of the Cuicatec ecosystem developed from historical evidence.

The Preclassic Period

Perdido Phase

Spencer (1982) and Redmond (1983) found a pattern of Perdido phase occupation on the first terrace above the alluvium bordering the Río Grande. On the basis of survey and excavation they interpreted these settlements as villages at a chiefly level of organization, based on irrigation along the low alluvium on both sides of the Río Grande.

My survey located one site from the Perdido phase across the river from Cuicatlán, in Valerio Trujano. This was a thin scatter of sherds on the first terrace above the Río Grande floodplain. There were no noticeable architectural remains. This is exactly the type of location described by Redmond (1983) for Perdido phase sites.

Lomas Phase

Spencer (1982) and Redmond (1983) describe a shift in settlement away from the first alluvial terrace to the piedmont spurs behind the first terrace above the Río Grande. This shift is accompanied by evidence of irrigation of the terrace with major canals, rather than the temporary canals used for irrigation on the bottom alluvium along the Río Grande. This shift does not seem to be accompanied by a growth in population.

Spencer and Redmond interpret this shift as the result of conquest of the Cañada by the Zapotec from the Valley of Oaxaca. In support of this is evidence of a different internal structure in the villages, the recovery of a skullrack from the site of La Coyotera near Dominguillo, major construction, and evidence of a population larger than could have been supported by local agriculture at Quiotepec.

The site of La Unión, near Atlatlauca, belonged to the Lomas phase occupation. This site was located well back from the Río de las Vueltas and sat astride the present road from Atlatlauca to the Valley of Oaxaca. I also found two Lomas phase sherds on otherwise predominantly Postclassic sites, at Iglesia Vieja and at my Site 16 in Cuicatlán. These were in locations back from the edge of the alluvial terrace, in locales like those described by Redmond as typical of Lomas phase sites.

The Classic Period Trujano Phase

In the Trujano phase Redmond (1983) believes that the Zapotec state withdrew from direct domination of the Cañada. As a result, there is a change in the structure of the Trujano phase settlements. Quiotepec becomes much smaller, and there is a gradual increase in population throughout the Cañada. I found a few Classic sherds in otherwise Postclassic sites in my survey. This agrees with Redmond's interpretation that Trujano phase sites were based on irrigation of both the high and low alluvium.

The Postclassic Iglesia Vieja Phase

Almost all of the sites in the Cañada are covered with Postclassic material. These sites are common and widespread. Characteristically, the sites are high up on the talus slopes or hill slopes that stand above the Cañada floor. I found no late sites down near the river on relatively flat land. There seemed to be no late sites in the terrain irrigated by the present irrigation systems, while almost all high slopes above the highest canals had remains on them. For instance, in Cuicatlán, I found two concentrations of sherds and remains right at the northern end of the irrigation system. Here, the mountains close in on the Río Grande, and there is no flat land on which irrigation could be practiced. The first of these had at least two house foundations. The second was a surface pottery concentration. The pottery collected was the usual gray and cream Postclassic forms, with some comales and one or two pieces of "Huitzo Red on Cream" (Hopkins 1973).

South of these sites, there is an area of flat land between the upper irrigation canal and the talus slope at the foot of the cliffs that bound the Cañada floor. There are no remains on this flat land or on the very steep talus slopes. The next major area of occupation remains is on the meseta that reaches out across the Cañada floor towards the Río

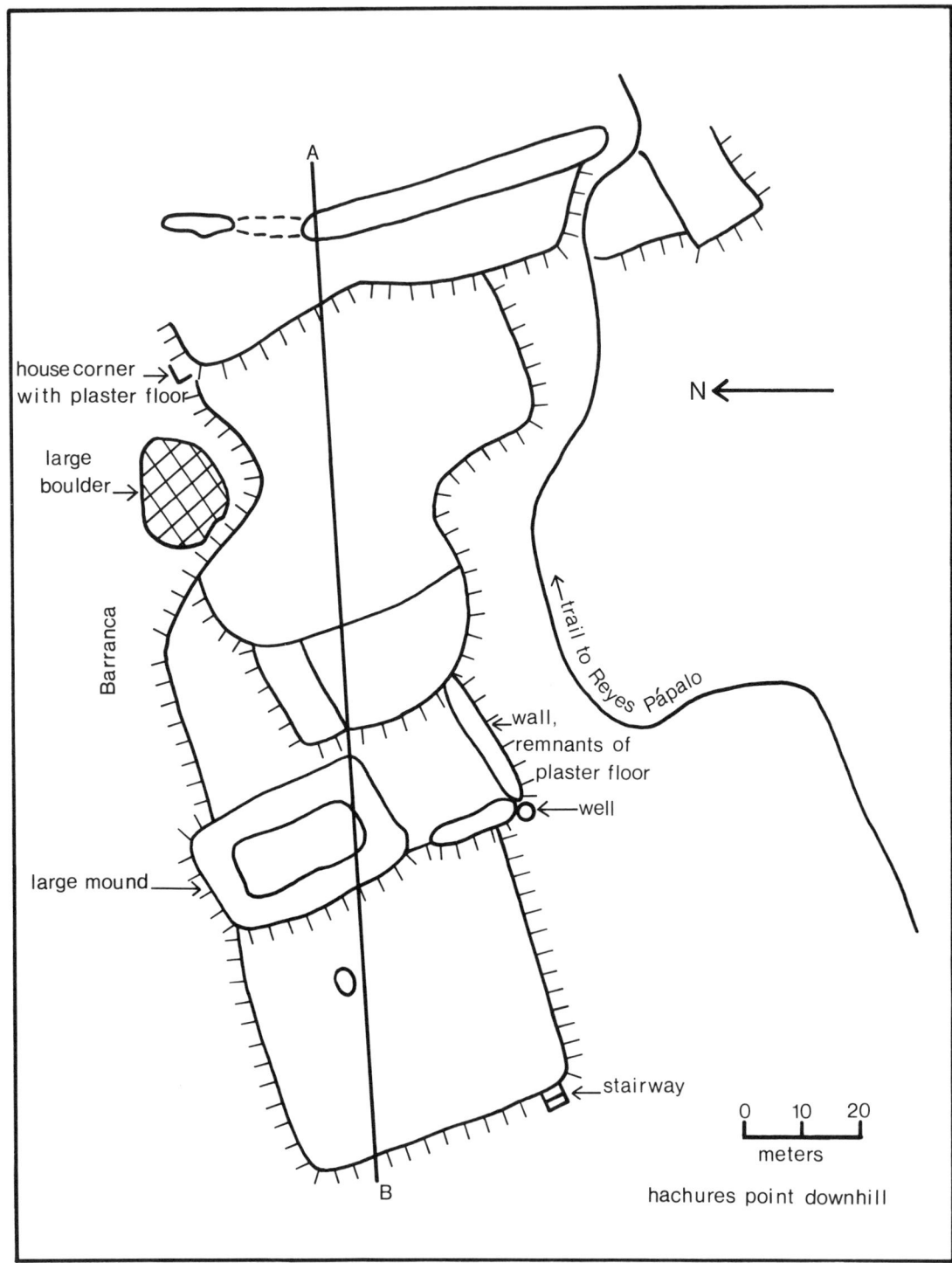

Figure 9. Ojo de Agua (Site 21).

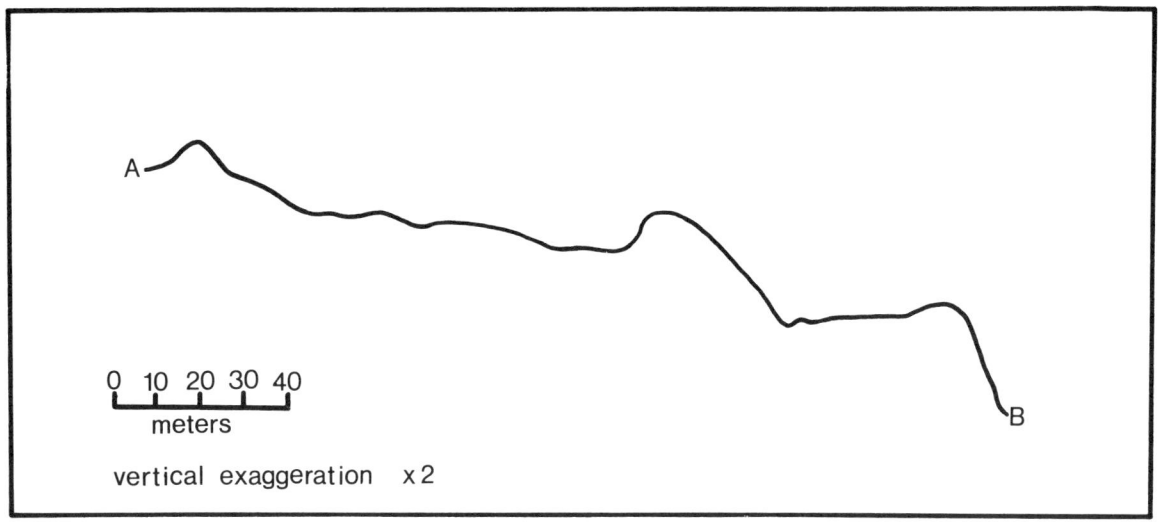

Figure 10. Profile A-B, Ojo de Agua (Site 21).

Grande, paralleling the course of the Río Chiquito. This meseta, a flat-topped ridge, is shaped like a boot, like a map of Italy. It rises some 115 m from the Cañada floor. At its upper end, it abuts the cliff that rises to the north-northwest of the canyon of the Río Grande.

The meseta is entirely covered with dense remains. Most of these are small rectangular foundations of houses, delineated by rows of stones and occasionally with plaster floors. In addition, there are at least two larger complexes. One of these included a large mound, some 6 m high. Looters used a bulldozer to cut a trench through the middle of this mound. In front of the mound is a circular earthwork, about 36 m in diameter, and two smaller mounds which stand about 1.5 m high. This is located in the middle of the meseta. Further down, on the end of the meseta where the present air strip has been bulldozed, is a probable "juego de pelota" or ballcourt. The two parallel banks of this feature have been cut by the air field, and one of the mounds now supports a wind sock (Fig. 13).

All the little hills that rise steeply from the irrigated fields to the south of this meseta are also covered with ruins. Both the small outlier just to the north of the present bridge across the Río Chiquito and the spur on which Cuicatlán lies are covered with house remains. The upper end of the spur, on which most of the present town of Cuicatlán stands, is the site called the Cerro de las Tres Cruces.

Another site, Ojo de Agua, sits across the trail to Reyes Pápalo, above the present town of Cuicatlán, but at the foot of the cliff that rises 200 to 300 m above and to the east of the town. One side of the site is bounded by a barranca. The site consists of a complex of terraces on the steep talus slope at the foot of the cliff (Figs. 9, 10). At the lowest part of the site is a large terrace platform, 45 m wide and 40 m deep. Above this a mound rises 12 m from the surface of the platform. Traces of reddish painted stucco can be seen. At the lower edge of the platform were the remnants of a stairway 3 m wide. There were other house foundations and plaster floors, as well as stone walls. At the top of the site is a long bank, some 3.5 m high and 80 m long, that the trail cuts through. The purpose of this wall is problematic. It could conceivably be a defensive work protecting the trail to Reyes. Another anomalous feature is a stone-lined circular shaft that looks like a well at one corner of the terrace, to one side of the mound and platform.

At the top of the cliff and paralleling the road to Reyes, along the crest of a ridge between the cliff above Cuicatlán and the less steep, but still precipitous drop to the Río Chiquito, is another group of large ruins, although these are badly destroyed. This site included one mound and a wall-enclosed compound. The mound is about 25 m x 35 m. The site stretches almost a kilometer along the ridge.

The collections from these two sites, the one directly above the other, yielded a number of unusual objects in an otherwise typical Postclassic collection. From Site 10, on the top of the cliff, in addition to the usual material, there were petate-impressed sherds, two of the few pieces of "Mixtec polychrome" we found, and a lifesize tan clay replica of a finger, broken off a larger piece. There also was one metate fragment from the site on the top of the cliff.

In addition to the regular Postclassic material, the flat, rectangular, petate-impressed "tiles," some 1.5 to 2 cm thick and roughly 35 by 20 cm, as well as the gray tubes, came from Site 21, at the foot of the cliff at Ojo de Agua. All of the tubes came from this site, or from a house construction site just below it.

These sites, from their location, may have formed the central part of the pre-Hispanic town of Cuicatlán, or at least one major barrio. The plaza, and related stairs and mound, show that at least part of this site represented some kind of ceremonial or public architecture. The defensive possibilities of the sites on the top of the cliff are obvious, since the trail must wind up a narrow ledge, but no definite defensive works were observed. The material from the sites gives an impression of greater richness. The sites had at least some ceremonial importance. The fragment of a metate and the dense cultural material would seem to indicate that at least part of the site on the top of the cliff was occupied. Of course, since this was on the road to Reyes Pápalo, the metate could also represent the remains of the trade going up the mountains.

From Ojo de Agua south along the talus slope at the foot of the cliff behind Cuicatlán, towards San Pedro Chicozapotes, the ruins continue. High on the talus slopes are a series of remains of house foundations and larger areas bounded by terrace walls and dotted with mounds. Plaster floors, and eroded-out or sacked tombs occur. Sherds are quite dense. The sites are continuous except where the talus slope is cut at intervals by barrancas.

I made one small excavation in one of these sites on the talus slope. This was Site 16 (Fig. 14). It was located on the talus slope to the south of Cuicatlán, just above the chapel at La Carbonera. I began this excavation with two goals. One was to obtain some samples suitable for flotation and recovery of plant material. The second, hopefully, was to get a stratigraphic sequence dating back to earlier times than the Postclassic surface material, since a few Classic sherds had come from this site. On opening the site, I immediately encountered a plaster floor, with a sort of step in it, rising about 7 cm. At the lower end of the site, where the plaster floor came almost to the surface and was eroded considerably, were two relatively complete pots, one a kind of gray double vessel; and the other a complete, but smashed, polychrome tripod olla. Between these was a sort of gutter-like feature composed of flat rocks standing on end, aligned with the step on the plaster floor. This feature was perhaps a meter long within the excavation and about 8 cm wide. Underneath the plaster floor and oriented with it was another plaster floor, which curved around a now nonexistent wall and was bounded by the step. This suggests that the first plaster floor (the highest one) was a reconstruction over an earlier floor. Beneath this was a level of burned rock that contained lots of charcoal and animal bone.

The preliminary report on a flotation sample taken from this level indicated the presence of pine among the charcoal fragments. Pine must have come from the highland communities, either as beams or boards for construction, or, more likely, as charcoal. This proves that a pattern of symbiosis existed in pre-Hispanic times.

Beneath this level was a terre pisé wall bounded on both sides by plaster floors which curved up to meet the wall. The wall was preserved to a height of some 30 cm. Two horizontal layers were detected in its construction. Included in the adobe of the wall were rocks 10 to 15 cm in diameter.

There were at least two separate building periods represented in this one pit, as well as apparently a remodeling of the second phase of building. The first was that associated with the terre pisé wall, while the second is represented by the two higher plaster floors. Disappointingly, sherds from underneath the lowest of the building levels included such late types as Huitzo Red on Cream.

Across the river from Cuicatlán in Valerio Trujano, the pattern of Postclassic occupation seemed to be similar. All of the low hills that stand above the irrigated fields were covered with remains of Postclassic habitations and sherd scatters. A more cursory survey recovered the same heavy occupation on the hills of El Chilar and Dominguillo. In general, one can find Postclassic remains simply by looking for an eminence above the irrigable land. One possible exception was a small Postclassic mound across the river from Dominguillo, on a

ARCHAEOLOGICAL EVIDENCE

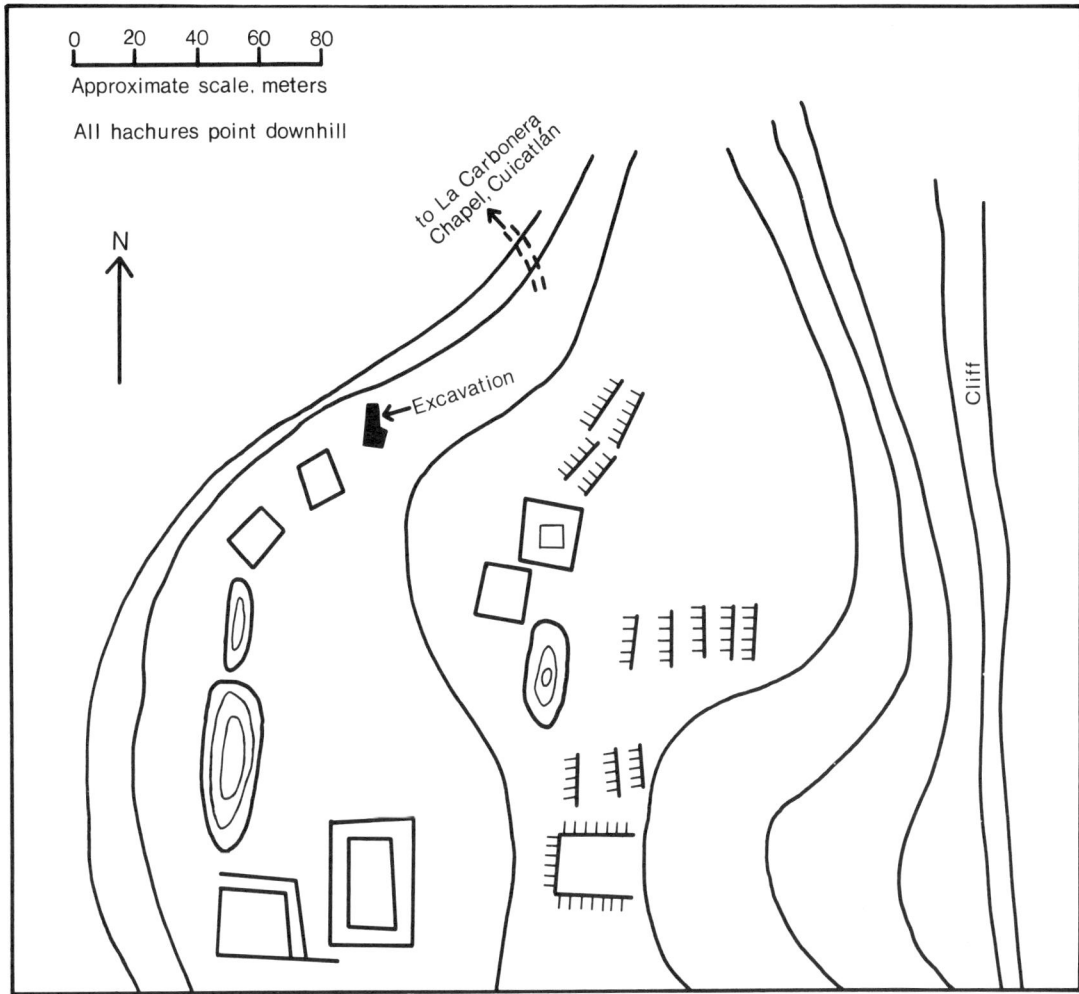

Figure 11. Site 16, Cuicatlán.

neck of land surrounded on three sides by the river. At present this side of the river is not irrigated.

As important as where remains occurred is perhaps where they did not occur. I made several attempts both in Cuicatlán and on the general survey of the Cañada to locate sites down in the low-lying irrigated area. There was always a very thin scatter of small Postclassic sherds in the fields. However, nothing that could be called a concentration was located, with the exception of the one Perdido phase site at Valerio Trujano and the Lomas phase site of La Unión, across the river from Atlatlauca. Both of these were located on the first river terrace.

Surveys on the fields of Cuicatlán revealed only a very thin scatter of sherds. Subjectively, these seemed to be broken into smaller pieces, which is probably a function of the land being turned over by cultivation. I identified one possible site in the irrigated area of Cuicatlán from a pattern observed on the aerial photos, but field inspection yielded pieces of roof tile and glazed pottery which indicated some kind of post-Hispanic feature.

Attempts to find material below the first river terrace, on the present river floodplain, were entirely fruitless. Apparently this material is turned over frequently enough by the river so that all remains have been carried off or destroyed.

This impression that the Postclassic material was not found down in the irrigated fields was reinforced by examination of trenches cut by public

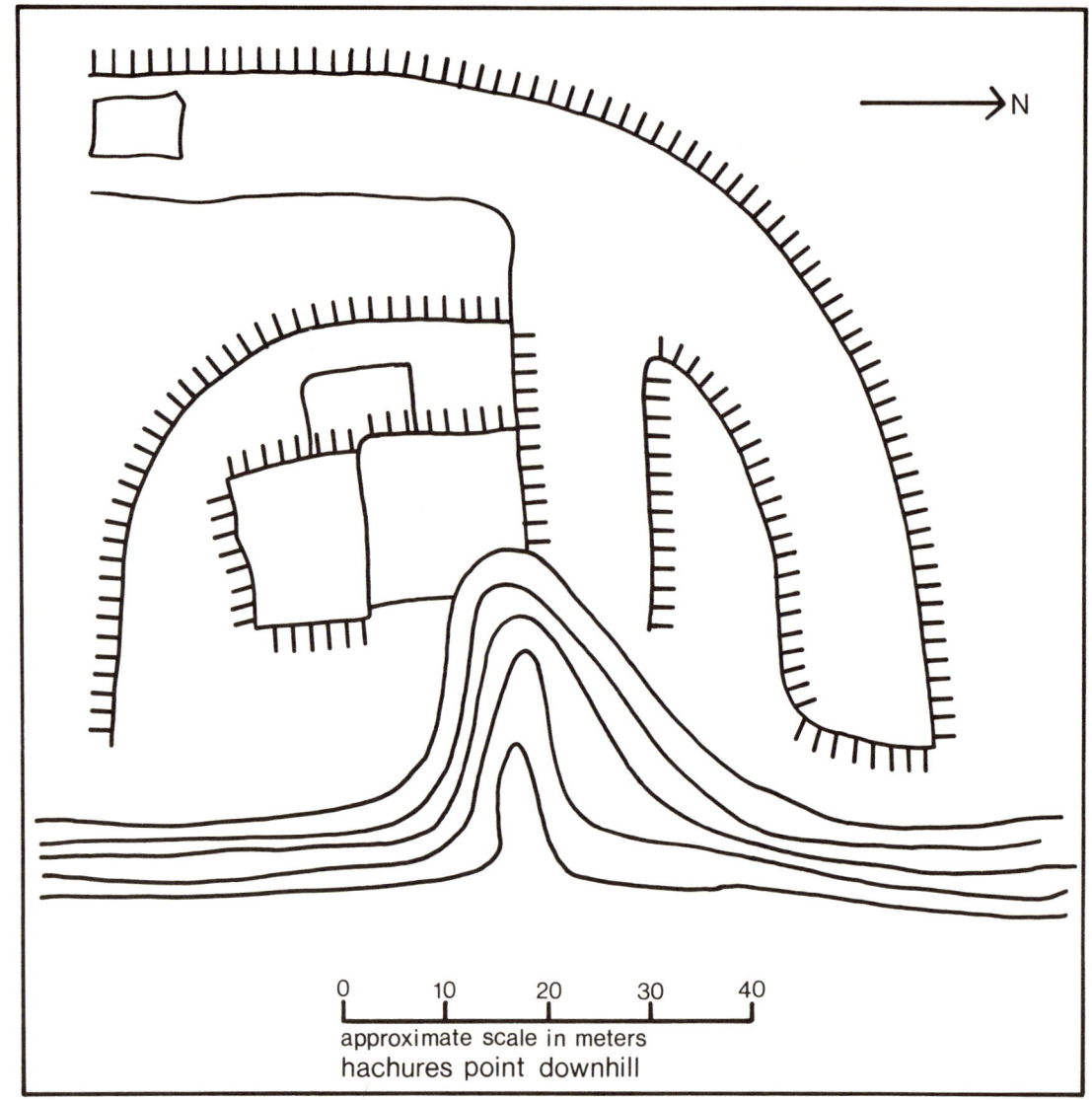

Figure 12. Site 18, Cuicatlán.

works in the streets in the lower part of the present town of Cuicatlán. Examination of these cuts and of the dirt removed from them and interrogation of the diggers indicated that archaeological material did not come from the lower part of the town.

Ancient Irrigation Remnants

Since the major interest of this research was that of agricultural systems and their remains, one of the major kinds of archaeological features that I sought and found were remains of old irrigation systems. The primary method for locating these was by walking up the various Ríos Chiquitos, and looking for remains. In addition, I asked the local people if they knew of such remains. Since they were engaged in irrigation, they recognized these features when they had seen them.

The first of these irrigation remnants was that known as the "Toma de los Chentiles" by the people of Cuicatlán, above the present *toma de agua* of the present irrigation canal. It is known by most citizens of Cuicatlán and recognized as an old canal by them. There was even a recent, unsuccessful attempt to revive the canal.

Figure 13. Juego de Pelota on Iglesia Vieja meseta, cut by air strip.

These remains are constructed of rock set in mortar. The best-preserved fragments are near the upper end of the canal, below where the *toma de agua* for this canal must have been. These fragments are built into a hollow in the cliff above the stream bed of the Río Chiquito (Fig. 15). By leveling up from the highest fragments of the canal bed, to the point where the river was level with the canal bed, I could establish roughly the *toma de agua*. Since the canal must have had a drop, this would be a conservative estimate of the location of the *toma de agua*. This point was about a kilometer upstream from the present *toma de agua* for that side of the river. Here the canyon for the Río Chiquito closed in completely, and the miniature floodplain with its one line of banana and lemon trees disappeared. The river filled the canyon from wall to wall at this point, even in the dry season.

One hundred and thirty meters downstream from the estimated position of the *toma de agua* is a small fragment of mortar and stones. Below this, another 25 to 30 m downstream is a larger fragment, some 3 m above the stream level and 22 m long. This fragment had remnants of the channel bed of the canal preserved. Another 100 m downstream was another fragment, which also had the

Figure 14. Building levels, Site 16. *Above left*, Floor 1, with step; *above right*, Floor 2, showing outline of corner of earlier wall; *below left*, relationship between Floor 1 and Floor 2; *below right*, terre pisé wall and Floors 3 and 4.

Figure 15. Toma de los Chentiles, on the Río Chiquito, Cuicatlán. *Above left*, Fragment 2 (upstream); *above right*, Fragment 3 (downstream); *below left*, view of cross section of Fragment 3, showing two levels of canal bed; *below right*, canyon of the Río Chiquito at point where old *toma de agua* would have been.

bed of the canal preserved. Leveling established that the canal bed dropped 1.10 m in the 100 m. At this point the bed of the canal was 5 m above the river bed. While this figure was not very exact, since it was measured with a hand level, it does confirm that the canal bed ran downhill. This drop of 1.1% agrees closely with the 1.3 to 3.2% (with one section at 0.7%) reported by Woodbury and Neely for the Xiquila canal in the Tehuacán Valley (Woodbury and Neely 1972:108).

At the same time, leveling established that the river fell 3.5 m in 100 m, for a gradient of 3.5%. This is comparable to that of the Xiquila River (Woodbury and Neely 1972:108) of 4%. The construction of the canal is also quite similar to that of the Xiquila aqueduct, except that the Cuicatlán aqueduct was constructed of rounded rocks that occur locally, while the Xiquila aqueduct was constructed of a local stone that fractured in rectangular shapes. Since the canal had a more gentle gradient than the river, the fragments on a line downstream every couple hundred meters rose higher and higher on the slope above the Río Chiquito (Fig. 16). Following these fragments, I came to a point where the canyon of the Río Chiquito opened out to the main part of the Cañada. Here the canal crossed a small spur that stuck out from the cliff that forms the east side of the Cañada. On this spur was a small site with the usual Postclassic pottery. The canal seemed to cut a gap through the neck of this spur, between the site and the rest of the spur. I hoped that this would allow me to date the canal.

Dating agricultural remnants like canals is quite difficult. One cannot date them by cultural material or even carbon found in them, since this could easily have been washed in from earlier sites uphill from them. The only way one can successfully date such a feature is to find a place in which its relationship to a more conventional site establishes its contemporaneity with the site. Then by dating this site one can establish the date of the agricultural feature. If the canal did run through the depression that defined one side of this site, then I had the means to date this canal. For this reason I decided to excavate this site.

I first put a trench across the depression that I believed was the point where the canal had crossed the neck of the spur. This was a trench 11 m long by 1 m wide, and almost 3 m deep at the deepest part. This trench cross-sectioned the canal, which did cross the spur here (Figs. 17, 18). I then excavated the site on the tip of the spur that the canal crossed. The site was very near the surface, and part of it had been destroyed by erosion. It turned out to consist of the remains of one house, 5 m wide by an indeterminate number of meters long (Fig. 21). The width could be established from the remnants of two of the walls of the house that were preserved, but the other two ends of the house had eroded away.

Between the two walls was a plaster floor, in relatively good condition considering that it lay no more than 15 cm below the surface and in part had been destroyed by erosion. Two meters from one wall and parallel to the two walls there was a definite step in the plaster floor (Fig. 23), rising some 7 cm. This step is almost exactly like that on the floor recovered from Site 15, except that the step in the house on Site 15 was underlain by rock, while that on this site was simply placed over dirt. Above the step the floor meets the surface and has been destroyed. There were four possible post holes that may represent where roof supports or wall supports stood. The remains of a large olla were found just outside the walls of the house. On the other side of the house, also just outside the walls of the house, was a concentration of gray ash, with a large number of sherds, in the shape of a shallow basin. The preliminary report on the flotation sample from this feature indicates that it included seeds, nutshells, and corn. This may have represented where a kitchen lean-to stood, outside the house.

I then put a trench through the plaster floor, to establish a profile of the site. Under the plaster floor, and oriented with it (roughly 30° west of north) were the remains of two tombs (Fig. 24), with a common wall between them. Only one course of the walls of the tombs remained, and they seemed to have been disturbed considerably before the house floor was constructed. I detected these tombs below the level of the floor from disturbed fill containing bone, charcoal, and some small sherds. However, it was extremely difficult to get a clear profile in the reddish, sandy soil that characterized all of this site.

As I can piece it together, the history of the site is as follows. First, the canal was put in. Then the two tombs were put in, probably nearly simultaneously, since they were built next to each other, with a common wall. Later the tombs were partly exposed by erosion, which removed one side of Tomb 2 and two sides of Tomb 1. The burials were

Figure 16. Downstream fragments of the Toma de los Chentiles, high on the slope above the Río Chiquito, Cuicatlán.

Figure 17. Cross section, canal of the Toma de los Chentiles, where it crosses Site 5.

not completely exposed, although Tomb 2 may have been more so, since the bones in it are more scattered. After this, the site was leveled, and the northeast wall was built to support the terrace on which the house and plaster floor were built. Since the plaster floor in part covers the fill on top of the eroded Tomb 1, this house was constructed after the tombs had been eroded, although it is possible that the existence of the tombs was known when the house was constructed. At any rate, the tombs, the house walls, and the house step all have the same orientation. This may be because the house was oriented with the pre-existing tomb. However, it seems more likely that both were oriented with the pre-existing canal. Since both the tomb and the later house are clearly from the Iglesia Vieja phase, we know that the canal must at least have been in existence at that time. It is impossible to prove how much earlier it was constructed. The reason for the location of this site alongside of the canal is problematic. The present-day canal, which runs around the spur on which the site sits, leaves the spur on a

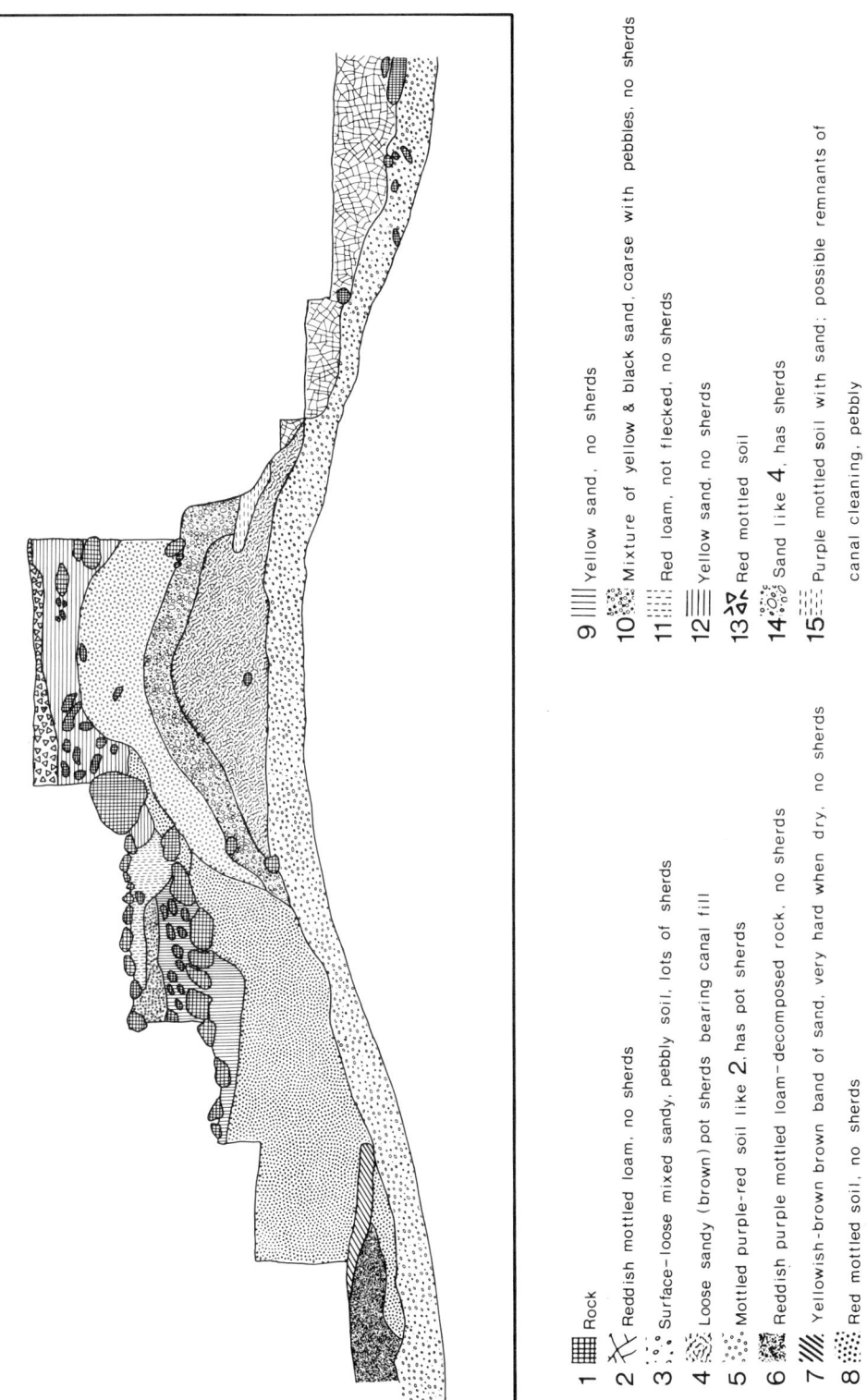

Figure 18. Cross section of trench cutting the canal of Toma de los Chentiles, Site 5.

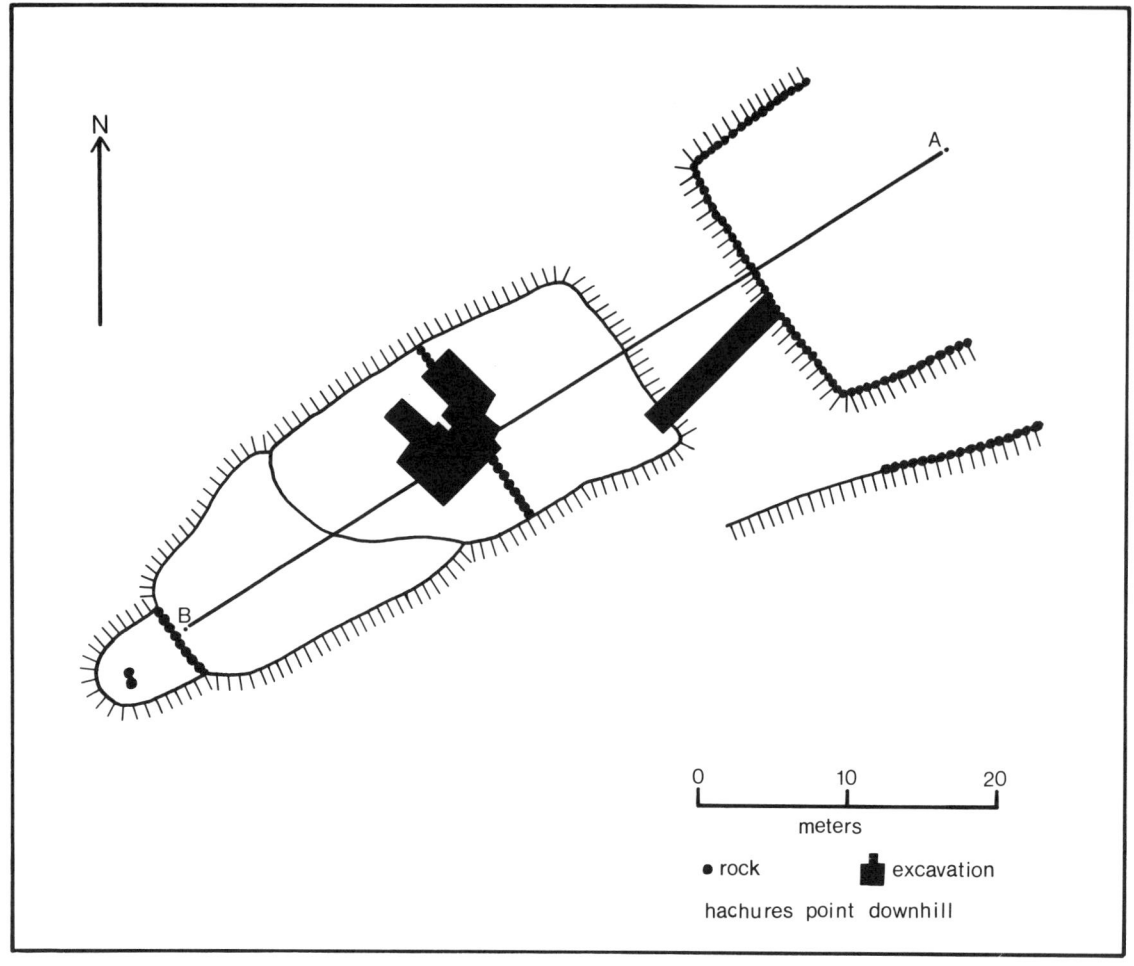

Figure 19. Plan of Site 5, showing excavations.

brick aqueduct which crosses the seasonally dry arroyo that divides the site from the meseta where the Iglesia Vieja site lies. The pre-Hispanic canal might also have used an aqueduct to get across the barranca, perhaps using *canoas* of hollow trees like those still used at Atlatlauca (Fig. 6). If so, this may have been a watch house to see that nothing happened to the *canoas*, since their failure would stop the flow for the entire canal. There may also have been some kind of diversion dam at this point, which was guarded, although I found no evidence for it.

After this point I lost the canal and have been unable to find remnants from here on. The barranca has undoubtedly removed the section of the canal just to the north of the site we excavated. However, I was unable to pick up further evidence of the canal to the north of the Iglesia Vieja, or on the other side of this meseta in La Rinconada, where the water of a higher canal would have been deployed. Consequently, the following must be regarded as speculation.

If the canal did continue beyond Site 5, it could have followed the contours of the Iglesia Vieja meseta around the entire length of the meseta, above and parallel to the route of the present canal. However, since it crosses Site 5 at an elevation of some 700 m above sea level, and it would have been not more than 500 m to the meseta, we would expect that the canal would, at its 1.1% drop, have dropped some 6 m in this distance. Now on the upper end of the meseta of the Iglesia Vieja there are two low gaps in the meseta, which are about 5 m lower than the point where we last find evidence

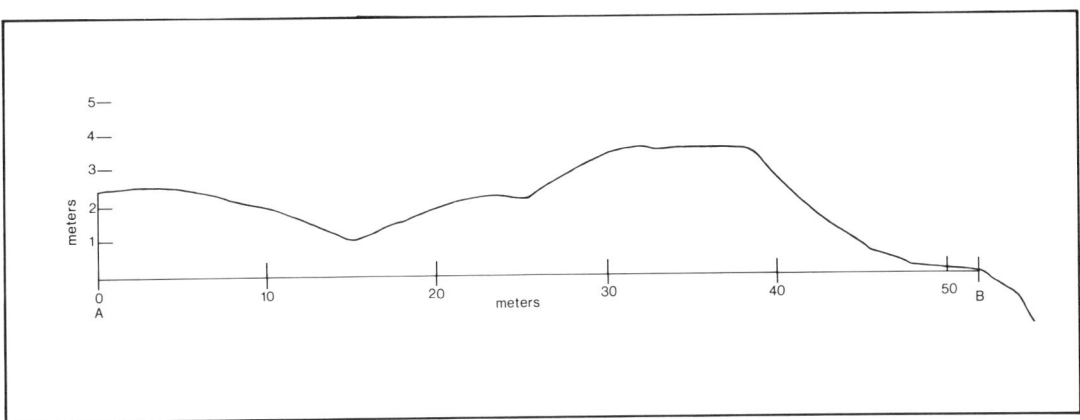

Figure 20. Profile of Site 5, Cuicatlán.

of the canal. These could have been where the canal cut through this neck. If the canal did in fact cut through at this point (and in retrospect I should have put trenches across both of these points to establish this) it would have had important implications for the irrigation possibilities to the north of the meseta.

Up to Site 5 we have established that the canal is 35 m higher than the present canal. If the older canal continued around the meseta as the present canal did, one would expect it to have maintained essentially the same elevation differential. However, if the canal was able to shortcut this circumnavigation it would have had an important advantage. This shortcut would represent a savings of roughly 5 km. At an average slope of 1.1%, this would mean a drop of 44 m. If the drop was more than that (which it may have been, given the short distance over which I was able to measure the slope and the accuracy of my instrument), if it dropped at 3% like the upper range of Woodbury and Neely's Xiquila canal, then the drop would have been closer to 120 m. This much altitude would have been saved by the shortcut. At the same time, there probably was a branch of the canal that went around the meseta, from which water would have been taken for the lands below this section.

The importance of this gain in altitude can be seen when one surveys the lands that would be potentially irrigable that lie above the present canal and below the potential irrigation of the new canal. There is a large area known as "La Rinconada" immediately to the north of the Iglesia Vieja meseta, and then a corridor of relatively flat land which would have been potentially irrigable by such a canal. A conservative measure of the area this represents is some 72 ha. If we extend the canal to its maximum potential, there is another area of flat land at the upper end of the present irrigation system which would add another 62 ha for a total of 134 ha.

Now, as Eva Hunt (1972) pointed out, the La Sabana canal, which is the canal that feeds this part of the valley at present, does not have enough water to supply the end fields in the La Sabana in the dry season. One might ask, then, what would be the advantage to added land area, if there was insufficient water to allow it to be used. The answer lies in two factors. The first is that some of the present canal is used for sugar cane production and other crops which demand a large amount of water. Pre-Hispanic crops demanded much less. The second is that the canal certainly would have had enough water to allow a wet-season crop in this area, which would otherwise be impossible. Even if these were only farmed once a year, they would considerably increase annual production. The fact that people will grow one crop a year, if that is all they can get water for, is attested by the present fields at the northern end of the La Sabana region.

Unfortunately, I could find no evidence of remains of the canal on the talus slopes above these areas, at the foot of the steep cliffs that bound the area. Walking the floor of these areas and even walking gullies cut in these floors, in the hopes of finding canals in cross section, I found no remains.

I also located no concentrations of sherds or any-

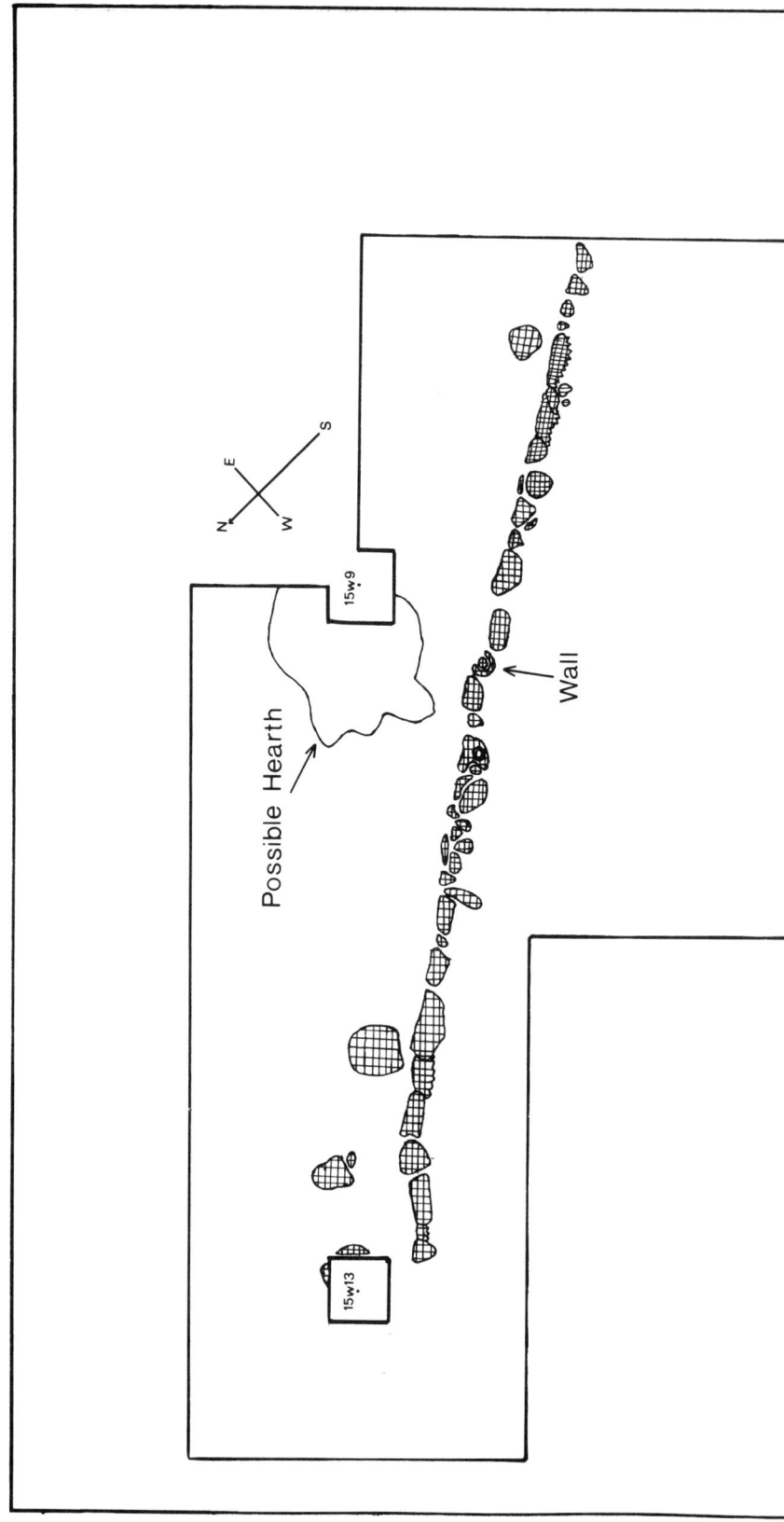

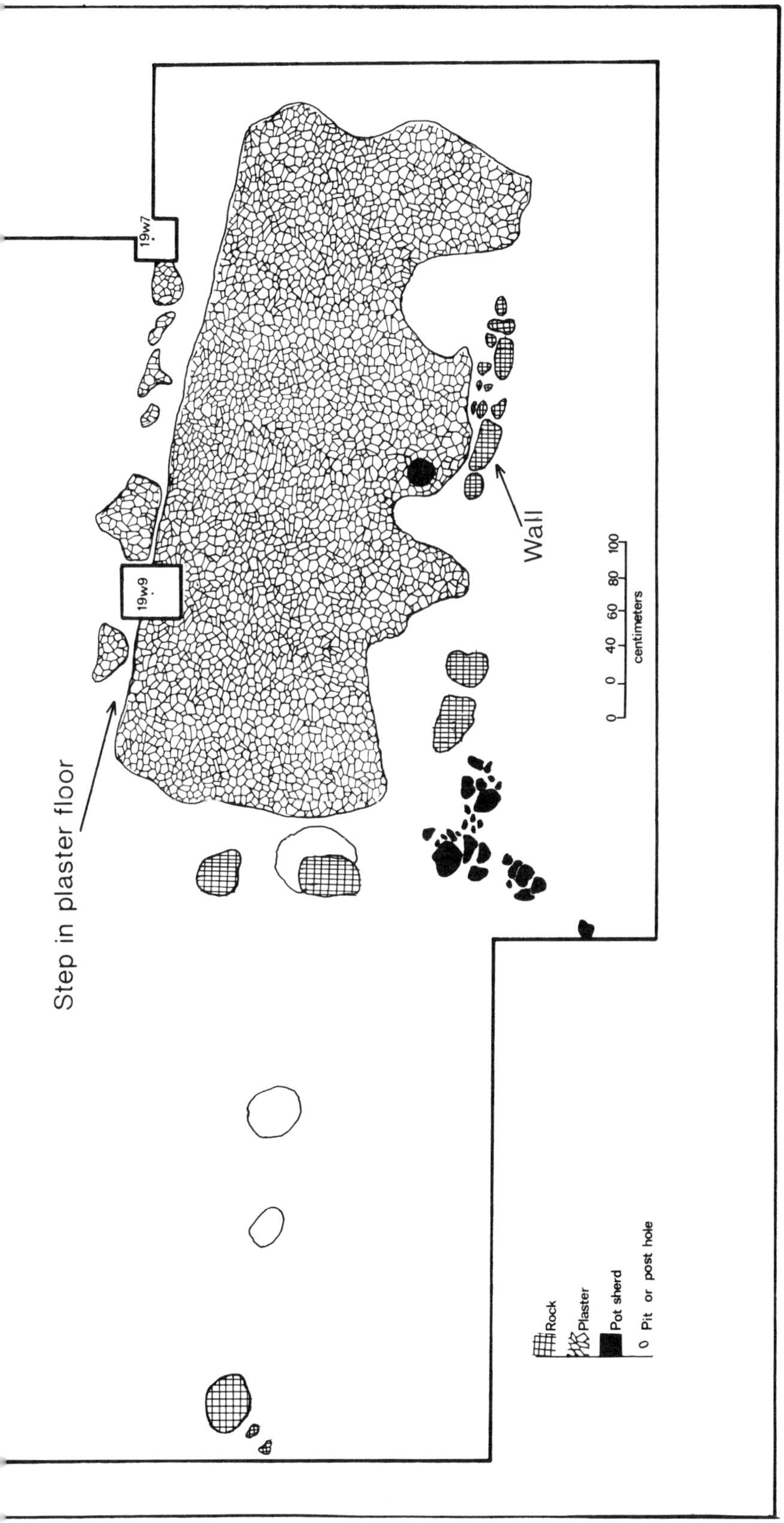

ARCHAEOLOGICAL EVIDENCE

Figure 21. Plan of house, Site 5.

Figure 22. Northeast wall of house, Site 5.

thing like a site in this area. Instead, the area was characterized by a very thin scatter of small sherds, like the scatter in the fields in use today, which I believe were always agricultural. This is in strong contrast to the heavy concentrations that are found above the irrigated area. I suggest that this scatter is similar because both areas were fields in pre-Hispanic times.

In addition to the Cuicatlán Toma de los Chentiles, which I was able to date, I found several other fragments on other, neighboring Ríos Chiquitos, which indicate earlier, higher *tomas de agua* on these rivers. In some other places I was unable to find evidence of higher canals.

One of the best-preserved canals was on the Río Cacahuatal. The Río Cacahuatal is the small river that joins the Río Grande between Cuicatlán and Quiotepec. At present it is considered part of the land of Quiotepec. The present *toma de agua* on the Río Cacahuatal is considerably downstream

Figure 23. Step in plaster floor of house, Site 5.

from the abandoned canal fragments, almost at the mouth of the river. As can be seen in the photos (Fig. 25), the remnants are well preserved and of the same rock and mortar construction built into the rock face, as the remnants at Cuicatlán and the Xiquila aqueduct. In this fragment, the side of the canal had not fallen away (although it was about to) and one could see the canal bed, which is about the same width as the Cuicatlán fragment, about 90 cm. Perhaps this narrow valley of the Cacahuatal,

which, unlike that of the Río Chiquito of Cuicatlán, did not open up to any major expanse of flat land, was all irrigated earlier. Today the main irrigation of Quiotepec is based on the other Río Chiquito, farther north, which opens up on a larger expanse of irrigable land. They are only now starting to expand into the Cacahuatal.

Across the river in Valerio Trujano, there are brick foundations for a bridge almost at the mouth of the Río Apoala, where it enters the Río Grande.

Figure 24. Tombs, Site 5. *Above*, Tomb 1; *below*, Tomb 2.

Figure 25. Abandoned canal fragment on the Río Cacahuatal, between Cuicatlán and Quiotepec.

Included in these brick foundations is a channel for a canal. This is now abandoned, as is the bridge, and the present-day canal is some 5 m lower at this point. Upstream from this bridge, and stuck to the rock face some 10 m higher than the bridge, is a small fragment of masonry and rock that looks like all the other pre-Hispanic canal fragments in construction, although the channel beds or sides are not preserved.

In this one spot we have evidence for successive lower levels of the irrigation system. The brick canal may represent a Colonial-period aqueduct, abandoned as the population continued falling, or an aqueduct from the turn of the century that was abandoned at the time of the Mexican Revolution. The arches of another abandoned brick aqueduct stand in the middle of the town of Valerio Trujano. The fragment of the mortar and rock canal undoubtedly represents a pre-Hispanic canal from a period when the water distribution system was even higher.

Between El Chilar and Dominguillo is the site of La Coyotera (see Spencer 1982). This site consists of foundations, a mound, and sherd concentrations on the top of a small hill standing up out of the valley floor. Behind this site and connecting it with another small hill is a stone wall about 2 or 3 m high at the highest, and about 3 or 5 m wide. There seems to have been an aqueduct between this small hill and the edge of the Cañada. Spencer (1982) dates this aqueduct to the Lomas Phase, based on sherds included in the construction fill of the aqueduct. Spencer also reports stone-lined canals intruding in, and therefore postdating, Perdido phase structures in Llano Perdido, near La Coyotera. Another similar aqueduct, which seemed quite a bit larger, was glimpsed above Dominguillo while returning from the site at Santa Cruz.

There were also places where I was unable to find remnants of earlier, higher canals, either because conditions had led to their destruction or because they never existed. At Dominguillo, the town takes its water off the Río de las Vueltas, not a Río Chiquito. I went up the Río de las Vueltas for several kilometers above the town, but did not find any remains. Not far above this town, this river, which is one of the major affluents of the Río Grande, fills its canyon bank to bank, so it is possible that either no higher aqueduct could be constructed with pre-Hispanic technology, or if one was constructed, its remains have been destroyed.

At Atlatlauca I also was unable to find any remnant of an abandoned canal above the present town on the Río Chiquito from which they drew their water. My impression was that the slopes were made of a material that, either because of its bedding or its composition, did not stand in the steep cliffs like the red sandstone and conglomerate in the lower parts of the Cañada. The slopes were active, and any fragment of the canal may have been carried away.

The irrigation system at Atlatlauca also impressed me as being less intensive than that of the lower parts of the Cañada. By this I mean that not as much of the flow of the two Ríos Chiquitos was used by the Atlatlauca canals. In addition, the canals were not so carefully constructed to keep the water dropping at the optimum slope while keeping the water as high on the slope as possible. At many points in the Atlatlauca system the canals descended miniature waterfalls, simply because some obstacle prevented the canals from continuing on a level.

It is possible that the present *toma de agua* is as high as it ever was, but that the earlier system used more of the water, and used it more carefully. However, I have no direct evidence for this from the short survey I did there.

In the historical analysis and ethnographic description, I suggested there was an uninhabited zone between the irrigated floor of the Cañada and the lowest field in the highlands. There are some sites not covered by the survey of the Cañada or reported in the literature that lie in this zone between the highlands and the lowlands. One of these is Santo Domingo (Bernal 1966). This site lies in the hills between Dominguillo and the canyon of the Tomellín, in land that is under the jurisdiction of the Mixtec town of Almoloyas. Bernal compares the architecture on the site to Mitla and Yagul. He reports all the ceramics found there are "Mixtec," which is to say, Late Postclassic. In addition to the architectural remains he mentions possible agricultural terraces. He does not see where water was obtained for the farming but suggests that there may have been some "aguajes entre los cerros" (Bernal 1966:9). Bernal did not have time to search for these in his short visit to the site.

Santa Cruz is another site located in this intermediate zone. It, too, is on the Mixtec slope above Dominguillo, and in territory belonging to Al-

Figure 26. Mineralized fragments of earlier canal beside present canal, Santa Cruz, Almoloyas, Oaxaca.

moloyas. This site is located just above the contact between the limestone that caps the Cañada and the reddish sandstones and conglomerates that form the bedrock of the lower parts of the Cañada. A small hamlet lies in the midst of a small irrigation system which is fed with water that comes out of a spring in a cave above the town. This stream has formed the Barranca del Aguacate, from which the *agua potable* of Dominguillo is taken. The town sits on a limestone slope above a considerable cliff. The water from the spring is led away from the barranca and deployed through a system of tanks and canals. The present canals are mineralizing rapidly from the carbonates in the water. Along the present small canals are remnants of older mineralized canals (Fig. 26), like lesser versions of those at Hierve el Agua (Neely 1967), and in the Valley of Tehuacán (Woodbury and Neely 1972). The slope and layout of these canals is like the "fossilized" canals described in the lower Valley of Tehuacán (Woodbury and Neely 1972).

In addition to the old canal and the tanks which have been revived, there are abandoned check dams in the barranca which are not now farmed. There is a grove of black zapotes, chicozapotes, and avocado trees which the inhabitants think may be the remnants of an earlier system. Above this town and above the irrigation system were considerable architectural ruins. I did not visit them because I was requested not to by the archaeological inspector. One of the natives of this town reported the remains of another small system of a mineralized canal, tank, and small expanse of land similar to Santa Cruz, but he refused to take us there because of fear of conflict with the Indian police of Almoloyas. Apparently the conflict between the Cañada Cuicatec communities and the Mixtec of Almoloyas is still active.

Another site, El Despoblado, lies in this same in-between zone, above the limestone contact, but considerably lower than the highland towns. The site is spectacular in its inaccessibility. It is on the Río de las Trancas, between Cotahuixtla and Dominguillo. The Río de las Trancas is one of the many small tributaries of the Río de las Vueltas just above Dominguillo.

Following this river down from Cotahuixtla, one first sees evidence of two canals, one on each side of the river, of the same masonry and rock construction as that of Cuicatlán (Fig. 27). Since these were up in the part where the water flows through limestone, they frequently are covered with deposits of carbonates. The remnants of one canal, on the south side of the river, could be followed on foot. On the northern side of the river, across from the canal, was a sheer cliff. Stuck in the middle of this cliff were fragments of another canal, at least 50 m above the nearest easily accessible point. This canal represented a difficult piece of construction.

Following both canals downstream, the canal disappeared on the southern side, where travel was possible. However, the northern canal ended in a long series of artificial agricultural terraces high above the river (Fig. 28). There are at least 26 terraces, around 3 m high, each one creating a field on a sheer mountainside. These terraces were in good condition. There were no architectural remains directly associated with these terraces, and a few Postclassic sherds were found on their surface. However, directly up the slope from these terraces on the ridge above them was a large site. This included many house foundations and considerable pottery, including some rather aberrant sherds. The pottery was Late Postclassic. At the highest point of this ridge is a large, well-preserved structure, with sloping walls with a complex profile (Fig. 29). The façade with its relatively well-preserved taludes and tableros rising in a complex silhouette seem closest in style to façades found at Mitla (Marquina 1951). This site is probably similar to the other sites of Santo Domingo and Santa Cruz.

Finally, a cursory investigation in Nacaltepec revealed extensive Late Postclassic sites on hilltops and the existence of abandoned terraces and check dams. There was, however, no way to date these. There are check dams buried in the *thalweg* of barrancas and exposed where the stream has been allowed to downcut in the barranca. These old check dams exist in barrancas that now have maguey *bancales* (terraces).

In summary, the few early sites found by my survey fit into the pattern for the Preclassic and Classic periods outlined by Spencer and Redmond (1983). The large Lomas phase site of La Unión suggests this pattern can be extended to the Atlatlauca arm of the Cañada. The position of La Unión, where the road from Telixtlahuaca descends to the Cañada, is interesting. The Middle Preclassic settlements were probably based on irrigation down in the alluvium of the Río Grande and the Río de las Vueltas. If Spencer and Redmond's convincing argument is correct, in the Late Preclassic Lomas phase, a Zapotec conquest necessitated the farming of the higher alluvium, using more elaborate and permanent canal systems. When the Zapotecs withdrew direct control in the Classic period, a population continued farming the high alluvium, and slowly increasing. However, by far the greatest number of sites represents, at the earliest, Late Classic and more certainly Early and Late Postclassic occupation. At this time there is evidence for heavy occupation in the Cañada in every place where irrigation systems were possible. Typically, the signs of heavy occupation are well up on the talus slopes or on the tops of hillocks that stood above the Cañada floor.

We also have considerable evidence that the irrigation systems were considerably more extensive, took water for irrigation higher, and reached higher lands than any system that exists at present. The evidence includes mortar canals and aqueducts that are located on lands that are now unirrigated. Those that can be dated are Postclassic.

In addition, we have evidence of sites in the zone between the highlands and the lowlands, which was nearly unutilized in historical times. In this zone, which is unoccupied today, there were irrigation systems in the small canyons of the Ríos Chiquitos, including one (El Despoblado) which had an elaborate system of agricultural terraces and a

Figure 27. Canal fragments on the Río de las Trancas, or Carrizal. *Above*, fragment from southern side of river; *below*, fragment on cliff on northern side of river that leads to El Despoblado.

Figure 28. Terraces fed by canal fragments on the north side of the Río de las Trancas, below El Despoblado.

canal that was a marvel of construction. Here, agricultural fields were sculptured out of the landscape.

The high location of the occupation sites in the Postclassic may represent an adaptation for defensive needs. However, they may also represent efforts not to waste potentially irrigable land for occupation. Certainly in Cuicatlán the lower boundary of occupation coincides with the present irrigation canal. Since the present irrigation canal on the town side probably is as high as it ever was, this is significant.

From the archaeological evidence sketched above, it is possible to discuss how the ecosystem took the form it had at the time of the Conquest, especially in the lowland towns. Secondly, we can then try to see which factors led to the establishment of this pattern.

In the Perdido phase, settlement consisted of small villages farming the alluvium along the major rivers, either by *humedad* techniques, planting where subsurface water could be reached, or irrigating by small canals which were redug each year. Floodplain irrigation such as this can be seen from the train at Parián, in the canyon of the Tomellín, en route to Oaxaca. Here, check dams like finger jetties are put along the sides of the floodplain at right angles to the direction of the river flow but not extending into the dry season channel. Canals are tiny trickles leading to fields planted right at

Figure 29. Large mound at El Despoblado.

the edge of the river. They usually are so small that diversion dams (to turn the water into the canals) are not used, or at most consist of a few rocks where the mouth of the canal opens to the river. In effect, the river is treated as a canal from which the field irrigation water is drawn at will.

In the succeeding Lomas phase, irrigation was introduced to the higher alluvium, not because of high population making it economical to farm these lands, but because of the artificial pressure of the Zapotec conquest. Initially, at least, this change was not accompanied by a large increase in population. If, as Spencer and Redmond argue, this new land was used primarily for production of specialty crops for export or tribute to Oaxaca like coyol and zapote, this deemphasis on staples may be one explanation for the slow growth of population.

In the Classic Trujano phase, Spencer and Redmond (Spencer 1982; Redmond 1983) argue that the direct Zapotec rule was withdrawn. However, settlements remained on the piedmont ridges up off the second alluvium, and the canal irrigation presumably continued, although Spencer and Redmond argue that production shifted to emphasize staples more than specialty crops. Gradually the population began to grow. The irrigation systems must have grown concomitantly. However, the Cañada towns must have played a low-key role, perhaps analogous to the Cañada at the time of the nadir of the post-Conquest population.

Only in the Late Classic and Postclassic did this region, and perhaps Mesoamerica, see a filling up of the landscape, making the production of the Cuicatec irrigation systems become again, for the first time since the Lomas phase production for the Zapotec state, a matter of regional concern.

Population, Irrigation Systems, and Settlements in the Cuicatlán Cañada

The largest settlements in the Cañada flowered in the Late Classic and Postclassic periods. They grew rapidly and seem to have been at their largest at the time of the Spanish Conquest. This occupation was based on irrigation systems much like those in use today. There is considerable evidence that these irrigation systems may have often taken off the water much higher on the Ríos Chiquitos than is the practice today, and they presumably would have been able to irrigate somewhat larger areas for at least one crop a year and possibly two.

Not only were the pre-Hispanic canals higher than the present ones, but the Postclassic occupation ruins indicate the value of irrigable land. Irrigable land was at such a premium that no one could afford to live on it. All the late sites uniformly lie above all land that could have been farmed. While defense may have played a part in the location of the Postclassic occupation zones, no identifiable defensive works were found that would support this interpretation for the pattern of occupation.

Further evidence of the desperate shortage of agricultural land and the lengths to which people went at the end of the Postclassic to grow crops is the existence of sites like Santo Domingo and El Despoblado, which were supported by systems of irrigated agricultural terraces. El Despoblado represents fields that were almost entirely made out of the hillside and a canal that must have been a phenomenon to construct and maintain, stuck as it is onto the face of a cliff at least a hundred meters high.

All of this indicates that in pre-Hispanic times the Cuicatec ecosystem had a higher population than it did when first recorded by the Spaniards, or than it does today. In addition to indications that there may have been more fields and more people in the lowland villages like Cuicatlán, there were sites in existence in the intermediate zone between the highlands and the lowlands. This zone seems to have been unoccupied from the time that Spanish records covered the area. These occupations represented agriculture so intensive that it begins to recall some of the Andean terrace systems, although on a smaller scale.

From the historical sources we constructed a model of the Cuicatec ecosystem and its workings over the time covered by the historical sources. We were able to isolate some broad factors that seemed to be important in the way the ecosystem worked.

First, differences in relief had a basic effect on the structure of the ecosystem. Because of the relief, the towns in the Cuicatec ecosystem were divided into two kinds, highland and lowland, in which different kinds of subsistence systems were possible, or necessary. Because of these differences a kind of complementarity existed between the two kinds of towns. Also because of the relief, the lowland Cañada towns lie on a major north-south, Mesoamerican travel route. Because of this, the Cañada towns were often on the interface of the contacts of the ecosystem with Mesoamerican macrosystems.

Population size seems to have been another very important factor in the working of the Cuicatec ecosystems during the historic period. The kind of subsistence and settlement characteristic of the Cañada depended on its guaranteed production being sold or traded to the highland towns that were within a few days' journey.

When population dropped, it hurt the Cañada towns most, since if there was plenty of land, with lower populations in the highlands, the markets disappeared for the intensive Cañada production. As a result, settlement in the Cañada became much smaller, and the irrigation systems were abandoned at their upper reaches, falling back into a core area. Only when the population began to rise again did the Cañada towns return again to their former importance.

Technology affected the Cañada first. The most important technological changes in historical times were improvements in the transportation through the Cañada, which served to highlight and even exaggerate the importance of the Cañada as a transportation route and as the interface with the macrosystem. Finally, from the beginning of the historical records on the Cuicatec ecosystem we saw that the factors that worked differentially on the parts of the Cuicatec ecosystem included the interaction of the ecosystem in which the larger macrosystem was immersed. This interaction with the macrosystem followed the lines of the structure of the ecosystem.

The archaeological evidence allows us to examine the processes that led to the foundation of the Cañada settlements. We can then see if the processes that operated in the pre-Hispanic system were the same as those in historic times or if some other factors existed that were more important then. The relief that differentiated the highland from the lowland towns initially led to a distinction between the areas that were utilized and those that were not. We do not know what the highland populations looked like in the Preclassic and Early Classic, but the lowland Cañada population was well beneath carrying capacity at this period.

One is tempted to explain the settlement of the Cañada by the introduction of the necessary irrigation technology. One could argue that the discovery and introduction of irrigation to the Cañada would explain the rise of the Cañada towns. An area that previously could not be utilized would become relatively important, because with irrigation, and its *tierra caliente* location, it could produce two crops a year. Only ignorance of the necessary technology earlier would have prevented its development.

However, for most of the history of Mesoamerica, irrigation techniques would have been known in the macrosystem around the Cuicatec ecosystem. Well before any major settlement of the Cañada, these techniques for irrigation existed within areas near the Cañada and probably would have been common knowledge. For the Valley of Oaxaca, Flannery et al. (1967) argue that a shift in the settlement pattern in the period of the Middle Formative, from 600 to 200 B.C., represents a point at which people began to rely in part on small irrigation systems with perennial streams on the upper edge of the piedmont zone of the Oaxaca Valley. Neely (1967) was able to excavate a mineralized example of such a site at Hierve el Agua and establish its date as pre-300 B.C. To the north of the Cañada there is also considerable evidence for irrigation techniques being employed before they were in the Cañada. Woodbury and Neely (1972) report a huge dam for storing water from an arroyo which dates from the Preclassic in Tehuacán. These would also almost certainly precede the Cañada irrigation systems.

We know, both from the ceramics and from the evidence of widespread long-distance trade from very early times, that there must have been contact through the Cañada and that knowledge of these irrigation techniques would have been easily available to the inhabitants of the Cañada. Conversely practitioners of irrigation in the valleys of Oaxaca and Tehuacán would have been aware of the existence of suitable land for the construction of irrigation systems in the Cañada. Consequently, we cannot invoke the lack of technological ability to explain the low population in the Cañada in Formative and Classic times. Irrigation was necessary for the importance of the Cañada, but it was not sufficient.

The historical evidence seemed to indicate that population density affected the Cañada towns considerably. When the population in the ecosystem in general declined, the Cañada towns declined disproportionately. We argued that this was because the Cañada irrigation systems depended on having a market for their produce in the neighboring highlands—areas that in off-years could not be certain of producing enough food. When population declined, these markets disappeared.

This pattern fits the arguments of Ester Boserup on the relationship between intensive agriculture and population. She points out that the less intensive agriculture is, the more efficient it is per man-hour (Boserup 1965). The advantages of irrigation systems and other forms of intensive agriculture is that they are more efficient per unit of land. This efficiency is accomplished, at least in preindustrial times, at the expense of an increased input of labor per unit of harvest. Thus, she argues, intensive agriculture will only be found when the population is high enough to force people to put more work into the system because the amount of land is limited.

This sequence seems to fit the historical processes of the Cañada. When the population declined and the pressure on the land declined, the decline was disproportionate on the intensive systems of the Cañada. And they grew back in importance when the population again grew and land began to be in short supply again. It seems, then, that population might be a factor in the importance of irrigation and the Cañada settlements.

The population of Mesoamerica was rising throughout the Classic period. At some time, the neighboring systems reached a point at which the irrigation systems in the Cañada took on new, regional importance. These systems, then seem to have prospered quickly, expanding the irrigation systems to the possible limits with the technique that existed.

Further evidence of the pressure of population making the intensive agriculture possible, and even necessary, is the existence of small but even more intensive systems like El Despoblado, where not only the canal, but the fields as well, were constructed by their users. Again, if we are correct about this relationship, then it is understandable that these systems fell into disuse after the Conquest, while the Cañada irrigation systems survived at a low level. We also have the evidence of check dams and terraces in the highland town of Nacaltepec, which seem to date from the Postclassic, that also represent more intensive agriculture than that practiced in the same area at present.

Finally, the occupation evidence seems to indicate a larger area was covered by occupation remains before the Conquest than that occupied at present. Admittedly it is difficult to calculate population from an area of remains, since one has to assume many variables such as the size of houses and the number of occupants per house. However, the short time period that characterizes the settlement remains means that at least the present and past are comparable in time of occupation.

The settlements that were begun in the Cañada seem to have been stratified. The two-class system described by the *relaciones* seems to be confirmed by the existence of remains indicating differentiation in the sites, such as the existence of various ceremonial centers, pyramids, etc. These are associated elsewhere in Mesoamerica with political and religious classes. Their size and fineness of finish imply the existence of administrators and full-time specialist artisans supported by the population. In a different way the old canals, of durable masonry construction, also seem to imply a higher level of expertise, at least in canal construction, than that used, at least on the local level, in the present system.

This stratification may have come about from the demands of the irrigation system. This has been argued to be the case, especially by Wittfogel (1955). However, as Lees (1973) points out in her summary, there are non-stratified societies that use irrigation systems. The relationship is not a rigid one. And, as Lees (1973) and Hunt (1972) both found, the local irrigation systems of both the Cañada and the Valley of Oaxaca were administered within the towns. Even where these systems involved several communities, they did not seem to demand particularly strong administration. At least there was a range of possible administrations. The irrigation systems of Cuicatlán were administered by not one, but four separate bodies (Hunt 1972; Hunt and Hunt 1973).

At the same time, it is apparent that part of the nature of the Cuicatec system was its relationship with the macrosystem. This is important on several levels. First, the macrosystem demanded extra work from the system to meet its demands. Secondly, it may have tended to enforce the existence of a hierarchical system, in order to acquire its tribute efficiently. In fact, irrigation of the high alluvium may well have begun as a result of direct intervention from the macrosystem.

From Cuicatlán and the Cuicatec evidence, it is clear that the Cañada irrigation systems played no part in the initial invention of irrigation in Mesoamerica. Both irrigation and civilization came into being elsewhere earlier than in the Cañada settlements. The Cuicatec Cañada irrigation systems were part of later Mesoamerican cultural developments. However, the form of the Postclassic Cuicatec ecosystem was part of a process that was taking place in Mesoamerica at a period long after the emergence of civilization. An increase in population in Mesoamerica did take place throughout its history. This increase was related to an increased reliance on intensive agriculture, as well as colonization of previously unoccupied regions.

Population and Intensive Agriculture in Mesoamerica

From the historical evidence I have argued that population density played a large part in the dynamics of the irrigated parts of the Cuicatec ecosystem. When the ecosystem lost population, the Cañada irrigated systems suffered disproportionately. They decreased enormously in importance. When population recovered, the Cañada towns resumed their former place of importance. From the archaeological evidence, I suggested that the location of sites on the tops of hills and high on the talus slope indicated a shortage of land and intensive agriculture to support a population that, locally at least, was probably bigger than the present population. The Cuicatec case represents one small part of a Mesoamerica-wide process of increasing population and increasing intensification of agriculture.

In the Valley of Oaxaca, Flannery et al. (1967) record a gradual increase in population through time as more and more farming techniques were brought into play, culminating in the highest population in the last period before the conquest.

> Furthermore, the four systems we postulate for the Monte Alban I period did not represent the final stage of Zapotec agriculture. In later periods (A.D. 100 to 900), large mound groups and habitation sites cover areas of high alluvium where only dry farming or flood water farming is possible. By A.D. 1300, the area farmed included not only the entire valley floor but also the lower slopes of the mountains, which were frequently terraced. The terraces apparently depended on rainfall, and on the lower evaporation rates of the north and east facing slopes. The data suggest that the assimilation of new farming techniques was also a way of bringing into cultivation more and more of the previously unproductive physiographic units of the valley: The greater the number of systems used, the greater the acreage producing at top capacity. [Flannery et al. 1967:453]

In another publication from the same project, Anne V. T. Kirkby (1973) attempted to correlate the distributions of archaeological sites with the potential productivity of the kinds of land utilized. Her work also shows a consistently rising population. Her evidence diverges from my data in interesting ways.

The first is that she shows an increase in population during the Preclassic, when the Valley of Oaxaca had a substantial occupation. Then the population densities leveled off through the Classic and had a later steep rise, starting around A.D. 900.

Secondly, she shows population density increasing to the present, while the Cañada evidence shows that the pre-Hispanic population declined after the time of the Conquest and even now may not have fully recovered (Kirkby 1973:143, Fig. 55). However, she still documents a rising population up to the Spanish Conquest.

For the Mixteca Alta, to the west of the Cuicatec area Spores (1969, 1972) records a population constantly increasing starting in the Preclassic period. He says of the early Ramos phase:

> A marked increase in the number of sites is noted for the Ramos phase which is believed to have begun around 200 B.C. and to have lasted until A.D. 250 or 300. Population was at least double that of earlier times, and it can be assumed that increased pressure was being exerted against the productive potential of the Valley. [Spores 1969:560]

In the next, Late Classic Las Flores phase, he says:

> Ninety-two sites were occupied, and this, coupled with the fact that unusually heavy and extensive deposits of Las Flores ceramics are found on 36 of these sites would indicate an upsurge in population. Not only are more sites being occupied, but there is also a new emphasis on larger centers, probably increased clustering of population, and a shift from earlier times in the choice of site location. [Spores 1969:560]

He notes that the population continued to increase in the Postclassic Natividad phase,

> From around A.D. 1000 to the Spanish conquest, the number and size of sites greatly increased. At least 111 of the 130 sites show definite signs of occupation, and 78 of them were intensively used. . . . The surviving Natividad architecture pales beside the impressive Las Flores structural complexes, but the very extensive accumulations of pottery and scattered stone make it apparent that a massive expansion did occur. [Spores 1969:561-62]

So for this area too, there is a history of continuous population increase to the time of the Conquest.

The evidence from the Valley of Tehuacán shows an increase in population over time, but not an even one. The population increases took place in two major bursts, separated by a leveling off, or even a decline in population. The first major increase in population was in the Classic (Early Palo Blanco) period (150 B.C.-A.D. 300). MacNeish et al. (1972) report a thirteen-fold increase in area occupied by sites of this period over the earlier Santa María period and estimate the population at 20,000 to 30,000, compared to the Late Santa María population estimated at 5,000 to 8,000.

In the Late Palo Blanco period (A.D. 300 to A.D. 700), MacNeish suggests there was a leveling off or even a drop in population, based on a decrease in area occupied by sites definitely assignable to this period. This conclusion is clouded by a large number of sites that are possibly, but not certainly, also of this period, and by the fact that this period was slighted in the survey "because of the exigencies of time" (MacNeish et al. 1972:429). The sites in this period are associated with terraces and canals. The population estimate for this period is 15,000 to 25,000.

The transition to the Postclassic (Early Venta Salada) period is not a sharp one. There is some increase in population. The population for this period is estimated at 25,000 to 35,000 people. More important for this period is the appearance of the first settlement large enough to be called a city, or at least a larger town than any before. There is a great decrease in the number of single room dwellings, and a great increase in multi-room dwellings, which complicates population estimates.

The Late Venta Salada period (A.D. 1150 to the Spanish Conquest) shows the second really dramatic increase in population. The population may have tripled from the Early Venta Salada to the Late Venta Salada. MacNeish et al. (1972:473) suggest a population of $100,000 \pm 20,000$. Much of this population was concentrated in three, or possibly four, settlements that are so large and so diversified that they seem to be true cities.

The pattern of population increase in the Tehuacán Valley is interesting. The first great increase in population is roughly contemporaneous with the rise of the great highland cities in Teotihuacán and Monte Albán. There is a leveling off or decline in population in the Late Classic (Late Palo Blanco) period, which continues without any dramatic break into the Early Postclassic (Early Venta Salada) period. Finally, in the Late Venta Salada, there is a surge of population increase like that observed everywhere in Mesoamerica. The temporary lull in population increase through the Late Classic and Early Postclassic suggests that the Tehuacán Valley was part of a larger system at this time, perhaps in conjunction with Cholula, which grew during the Late Classic and Early Postclassic.

Farther afield, for the Valley of Mexico, Sanders (1972) summarizes his data for the Valley of Teotihuacán in a graph, showing a steep increase in population to the time of the fall of the city at the end of the Classic, an immediate drop, and then a quick increase in population to a level at least as high at the time of the Spanish Conquest as it had been in the Classic. The population curve for the neighboring Texcoco region shows a complementary pattern (Parsons 1971). The population declined through the Classic and then rose quickly after the fall of the city.

The net effect of the two together is of a constant rise in population. While Teotihuacán was growing and dominating the Valley of Mexico, there were temporary setbacks in the growth of population in the Texcoco region, as population was centralized in Teotihuacán. These were offset at the time of the fall of the city, as the population of Texcoco rose quickly with the influx of people (presumably from Teotihuacán). Sanders says of this: "I suspect that a graph representing the entire Basin of Mexico would reveal a virtually continuous upward curve of population from the middle formative period up to the time of the Spanish Conquest" (Sanders 1972:110).

Finally, Adams (1961) notes that the major population influx for the Chiapas highlands was in the Late Classic period. This population seems to have continued through the time of the fall of the Classic civilizations, without a break. Later, after the Early and Middle Postclassic, there was a major change and relocation of the population centers. However, Adams argues that this does not represent a reduction of population but rather a change in the settlement pattern and of the structure of the settlements: "In effect, a redistribution rather than a reduction of population seems to have taken place within the region as a whole" (Adams 1961:21).

While population was increasing in much of Mesoamerica, there is also evidence from the same areas of an increase in use of intensive agricultural techniques over time. For the Valley of Tehuacán, to the north of the Cañada, Woodbury and Neely (1972) document increasing remains of intensive agricultural systems over time, with the exception of the Purrón Dam, which was built in the Preclassic and abandoned in the Early Classic (Spencer 1979). However, the rest of the remains of various forms of intensive agriculture show a pattern that seems to be correlated with the rise of population previously mentioned. In the Late Classic (Palo Blanco) a trend of intensification of agriculture began which continued through the Venta Salada (Postclassic) period. The two small dams uncovered by the survey could not be dated very securely, but what evidence there was dated them to the late Palo Blanco or the Venta Salada period—that is, to the Late Classic and the Postclassic. The Xiquila aqueduct, a long masonry canal like those I located in the Cañada, could be dated more certainly by its close relationship with a number of sites. The aqueduct, which is actually two aqueducts, had an interesting relationship with the sites.

> The chronological aspects of the sites associated with the aqueducts are very significant, for most of the Venta Salada sites (TR 411, 405, and 412) are located upstream and could only be associated with the higher and longer aqueduct. They also may be oriented toward TR 296, a large fortified hilltop town of late Venta Salada times. The earlier sites, all dating from the late Palo Blanco subphase are located downstream along the Xiquila or immediately north of it along the Rio Salado and could be associated with either or both aqueducts.

One cannot help but speculate that the lower aqueduct was built in late Palo Blanco times and was used mainly during that period, whereas the upper aqueduct may have been built in that period but was used throughout the Venta Salada phase. [Woodbury and Neely 1972:109]

The terrace systems discovered in the course of the Tehuacán project show the same pattern of increase in use of terraces through time. The dating of these terraces is based in some cases on simple identification of sherds found on the terraces, which is not necessarily a reliable indicator of the date of the terraces. However, some of the sherds used for dating the terraces were from what were apparently field houses associated with the terraces. These would be more reliable indicators of the date of the features. Woodbury and Neely (1972) report three terrace sites from the early Palo Blanco subphase (Early Classic), ten sites from the late Palo Blanco (Late Classic), five from the early Venta Salada, and nine sites of the late Venta Salada (Early and Late Postclassic). Since the Venta Salada period is much shorter than the Palo Blanco, this indicates a definite increase in the hillside terraces over this time. Of the valley terrace systems (they distinguished hillside terraces from what are sometimes, as they put it, "misleadingly called check dams"), four are dated to late Palo Blanco, seven to the early Venta Salada, and ten to the late Venta Salada phase. Again, the dating was done on the basis of sherds collected from the fields, without other associations, and as such is not necessarily reliable.

Woodbury and Neely summarize the relationship of the valley and hillside systems as follows:

Thus from a start in early Palo Blanco times, hillside farming systems show a continued and substantial use to modern times. There are no valley systems as early, and in late Palo Blanco times valley systems are only one-half as numerous as hillside systems. In Venta Salada times the two types are about equal in number. [Woodbury and Neely 1972:116]

They attempt to explain the later development of the valley systems:

Although, as has been pointed out earlier, valley and hillside systems are only variations on a single principle of water control, it may be significant that the technique was developed first for hillside use and later extended to the valleys or small intermittent drainages below and between the hill flanks. Although greater amounts of water might be collected in some of the valley locations, the flanks may well have been more desirable for farming from the standpoint of freedom from flood damage and simplicity of terrace construction. [Woodbury and Neely 1972:125]

In other words, valley systems may have represented more work than the hillside systems.

Woodbury and Neely specifically link these terrace systems to increasing population:

The increase in the use of lines or walls of boulders to border and support fields may have come from a growing appreciation of the effectiveness of such systems and also from the demand for more and more land for cultivation, as the villages and seasonal camps of the Tehuacán Valley evolved to a growing number of secular centers deserving the term town or city. An estimate of population to be published in volume 5 of this series suggests a four fold increase from Palo Blanco to Venta Salada times. Concurrently, with the increase in population there was a growth in commerce and in concentration of political power, both trends requiring an expanding agricultural base....[The smaller size of the plots in the later valley systems than the hillside systems] can perhaps be understood best as a reflection of the relative scarcity of arable land with adequate and manageable water supplies in the small valleys of intermittent streams. [Woodbury and Neely 1972:125]

The other form of evidence of intensive agriculture in the Tehuacán Valley, the "fossilized" canals, also were dated, by indirect lines of evidence, to the late Palo Blanco and Venta Salada period.

The evidence for the Oaxaca Valley is somewhat different, in that the rise in population and the incorporation of intensive techniques seems to have taken place at an earlier time. Thus Flannery et al. report evidence for both canal and pot irrigation in the Formative (Flannery et al. 1967; Neely 1967). Anne Kirkby's analysis of the population of the valley and the land occupied shows the establishment of canal and pot irrigation between 1000 and 300 B.C. (Kirkby 1973). The land newly occupied for agricultural purposes at this time seems to indicate that these intensive agricultural techniques were in use. The lands added to the system after that time can be explained by the advantages accrued through time from the improvements in the productivity of different varieties of corn.

Thus, intensive agriculture seems to have been practiced much earlier in Oaxaca than in Tehuacán. This is apparently because Oaxaca was a nuclear center in which one of the first civilizations arose. At the same time, there is evidence for a constant expansion of the Valley of Oaxaca's production, through more productive varieties of corn, and continued expansion of the application of intensive techniques through time.

An increase in the intensification of agriculture through time paralleled the rise in population for the Mixteca Alta, to the west of the Cañada. For the Las Flores (Classic) phase, Spores says:

> The number, size and apparent intensity of utilization of the sites suggest that the valley was beginning to fill up and that there may have been considerable pressure being exerted against the productive resources of the valley. [Spores 1972:185]

One effect of this increase of population was a change in the area in which settlement was located, in order to open up farm land.

> The valley floor continued to be intensively farmed, but many settlements were moved onto higher elevations. The desire to open additional farmland along the formerly occupied lomas and piedmonts may well have furnished the impetus for the shift in settlement. [Spores 1969:560-61]

In addition he cites hillside terraces adjacent to Las Flores sites, which may have been partly agricultural.

In the later, Natividad (Postclassic) period, Spores says the *lama-bordo* technique began to be used, in which hillsides were deliberately exposed to erosion to supply fresh soil for fertilization to downhill terrace systems.

> This is the period of maximum Lama-Bordo terracing, a major technological and adaptive innovation, which we have interpreted as a mechanism aimed at increasing agricultural production in response to increasing needs of a growing population, growing demands of a tribute-collecting royal-noble elite, and the superimposed tributary demands of the Culhua-Mexica Empire. [Spores 1972:189-90]

> Unusual demographic pressures in the Nochixtlan Valley (in the Postclassic) made it necessary to use all available bottom land that was not occupied. However, even this appears to have been inadequate to meet the needs of a rapidly expanding population. But without substantial improvement in agricultural technology based on hand labor, the use of the digging stick, and apparently almost total dependency on rainfall and some diversion irrigation no substantial increase in productivity could be expected.
>
> The Mixtecs of the valley responded to these needs by constructing stone and rubble dikes designed to trap water and eroding soils as they descended the natural drainage channels that extended from the mountain to the valley floor during the period of heavy summer runoff. [Spores 1969:563]

The Valley of Mexico is intermediate between the pattern of the Valley of Oaxaca and those of Tehuacán and the Mixteca Alta. There is some evidence, none very clear, of early irrigation there (Sanders 1956; Millon 1957). There is also considerable evidence for increasingly intensive agriculture through time. A system of terraces along the southern edge of the valley, examined in 1966 by the author (Hopkins 1968a) revealed solely Postclassic sherd scatters, although admittedly this is not an entirely satisfactory means of dating the terraces. The chinampas which they bordered represented perhaps the most intensive and productive agricultural system found in Mesoamerica. While Armillas (1971) records two possible chinampa sites from the Preclassic, and possible chinampa-using sites on the shores of the lake basin in the Classic, he dates the major development of the chinampa to the Late Postclassic. Parsons (1971) suggests that a number of intensive agricultural projects in the Texcoco region were Aztec constructions. This included possible canal irrigation, floodwater irrigation, large drainage projects on the lakeshore plain, and terracing. This period is also the period in which two large aqueducts were constructed that led water from springs to the famous Baños de Netzahualcoyotl (Parsons 1971), although Parsons questions whether this system was used substantially, or at all, for agricultural purposes. Since we know the Valley of Mexico was heavily populated at the time of the Conquest, both from direct historical statements and from the fact that it was the center for the expanding Aztec empire, it is no surprise to find that the agricultural resources of the valley were apparently exploited to a degree of intensity found nowhere else in Mesoamerica.

Finally, it seems possible that the Late Classic and Postclassic increase in settlement in the highlands of Chiapas may have been based on terrace agriculture. Adams says settlements in the Late Classic and Postclassic seem frequently to be located on hilltops for military reasons. He also describes some apparently dispersed settlements:

> A more common variant was suggested by thin scatterings of debris at intervals along low terraces that followed the intersection of interior basins or level shelves with slopes leading up to headland settlements (CV-5, 17, 19, 22, 23 and Moxviquil). From the absence of debris for considerable distances along such terraces it seems reasonable to conclude that their primary function was agricultural: they assist in erosion control, and they serve to impound runoff from the valley slopes above. However, since sherds do occur at intervals it is also possible that scattered hamlets or house sites were placed along them. . . . At any rate, while the real extent of this kind of agricultural terracing along the valley-margins is considerable in some cases (See especially that around the Aguacatenango Valley . . .) the density of occupational refuse is so slight as to suggest that the bulk of the population generally occupied the headland sites. This is especially true since the scattered debris might not represent former residences *in situ* but instead might have been carried out to agricultural terraces as fertilizer, a practice still followed in some of the Indian communities today. [Adams 1961:16]

In the Postclassic in Chiapas there was a shift in settlement to larger valleys. There is some evidence of possible irrigation.

> What distinguishes between late postclassic and earlier times is primarily the shift toward the larger valleys which could sustain more intensive patterns of exploitation, and toward

the formation of organized groups of communities, centered on a "capital." [Adams 1961:21]

Consequently, there seem to be two related trends that characterized the developments in Mesoamerica, especially from the Late Classic on. The first was a tendency in nuclear valleys and other settled areas in the highlands, for the demands of an increasing population to be met by turning more and more to intensive methods of agriculture, including terracing, floodwater and canal irrigation, and increased reliance on intensive systems like pot irrigation and chinampas. This intensification was most noticeable in the Late Postclassic, when every possible method seems to have been used in every place in which they were possible.

Another aspect of this process was a kind of filling-in of the zones in between the major "nuclear" valleys. Areas such as Tehuacán, the Mixteca Alta, and even Highland Chiapas are described by the researchers who worked on them as clearly not being "nuclear" or central to the history of Mesoamerica in the Classic period. Yet all these witnessed a quantum jump in population and settlement in the Classic and an even greater increase in the Postclassic. In part, this filling-in was also accompanied by intensive agriculture.

It is problematic as to what relationship this filling-in process had with the developments in the more nuclear areas. There are two possible ways of interpreting this. One is that the filling-in process represented the expansion of the nuclear centers, as another solution to the increasing population. The other is that this filling-in process represented natural growth, perhaps augmented by refugees who fled the nuclear valleys as they became increasingly strained.

There seems to be no evidence that the filling-in of the in-between areas represented a deliberate, planned colonization. The same process of rise in population that seemed to be resulting in increasing intensive agriculture in the central, nuclear valleys, was resulting in increasing population and ultimately intensive agriculture in the in-between areas. With the possible exception of Highland Chiapas, all seem to have been in some kind of contact with the nuclear center. In particular, in the Mixteca Alta, the change from the Classic to the Postclassic was marked by a change in settlement patterns that suggested a change from defensible positions during the Classic. This in turn suggests that the contact with the Classic nuclear centers of this area, at least, was probably less than amicable.

Another sign that the in-between areas were not under the direct sway of the highland nuclear centers is that the periodization of these areas is not tied very closely to that of the nuclear. For Chiapas, Adams notes that the Tsah period, which he describes as Late Classic, continued through into the Early Postclassic. In the text he refers to this period and its developments as "late Classic and early Postclassic." Similarly, for the Mixteca Alta, the Late Classic Las Flores phase is described by Spores as being "gradually phased out between 900 and 1100."

Now it is well known that the "Classic" period does not perfectly correlate throughout Mesoamerica. Research at Teotihuacán (Millon 1967) has shown that this Classic civilization began slightly earlier and ended a couple of centuries earlier than the Classic Maya centers on which the Classic was originally defined and dated. However, the evidence from the areas between the Classic highland centers indicates that the fall of the Classic centers had little immediate effect on the marginal areas. The in-between areas do not usually have sharp breaks in their sequences that correlate with the fall of the Classic centers.

On the other hand, there is real reason to say that these marginal areas were in some way tied to the developments in the highlands. There are changes in all of them that are related to the rising of the new Postclassic pattern. In Chiapas, even though it was clearly off the mainstream, there is a major shift from small centers on the slopes around small valleys, to major larger areas, which seems related to some change in the polities. Similarly in highland Mixteca Alta, Spores reports a shift in the look of the settlement in the Postclassic Natividad period that seems to represent a change in the relationship of the settlements to each other and to the main Mesoamerican centers. While the Nochixtlán Valley was definitely undergoing a major increase in population and in intensiveness of agricultural methods employed from the Late Classic to the Postclassic, the major architectural remains seem to have deteriorated. Spores says:

> Many mountain top settlements of the Las Flores (classic) phase, including several large centers were abandoned or only partially utilized, reflecting the pronounced tendency for the settlements to cluster once again along the piedmont, piedmont spurs, and lowlying ridges. The surviving Natividad architecture pales beside the impressive Las Flores

structural complexes, but the very extensive accumulations of pottery and scattered stone make it apparent that a massive expansion did occur. [Spores 1969:562]

Thus the evidence at hand shows that, predictably, the areas in between the highland Classic centers that began to fill in during the Late Classic period were at least partly affected by developments in the centers, but they were far from rigidly tied to them. The process of filling-in seems to have established relatively independent individual small "city-states" in between the larger, older valley systems. Finally these were in the process of being united as part of a larger polity by the Aztec when the Spaniards arrived.

Thus the Cañada irrigation system seems to have represented a small example of a major trend that was going on during the Late Classic. Population pressure seemed to be making it necessary for increasingly intensive methods to be employed in order to increase the agricultural production to support the population. Or something in the Postclassic civilization was encouraging the use of increasingly intensive agricultural methods and the increase of population made possible by this. At any rate, there came a time when the kind of intensive agriculture with irrigation that was the only way the Cañada could be farmed became essential since everywhere such intensive agriculture was necessary, although, as we have seen, not everywhere was it possible.

Once these conditions prevailed in Mesoamerica, the advantages of irrigation as a production system would have begun to assert themselves. The higher potential production and greater security of production of the irrigated fields of the Cañada towns would make them begin to increase in importance. Without the high population throughout Mesoamerica that made irrigation economical, it would never have begun. The advantage in production per unit of land is irrelevant if there is enough land to go around.

At the extreme of this trend towards intensification, the systems like El Despoblado were constructed, in which not only canals, but the fields themselves were constructed. Surely, these represent a final indicator of the increasing intensification.

Cuicatlán and Mesoamerica

The Cañada settlements seemed to have been one example of a general process that was going on in all of Mesoamerica. Beyond this, what role did the influences from the macrosystem play in the functioning of the Cañada towns, both in their initial colonization and their later development? We know that the Cañada settlements seemed to represent an example of the process of increasing reliance on intensive agricultural methods that was going on all over Mesoamerica. This process seems to have been in response to the increasing population that seemed to be characteristic of Mesoamerica from the Late Classic on. However, did the individual Cuicatec Cañada settlements relate to some larger system? Were these settlements in some way tied to and dependent on a system larger than that of each town with its own irrigation system?

At the lowest level, there is no question that the Cañada towns never existed as independent, self-sufficient entities, based on irrigation from the Ríos Chiquitos. The conditions that led to their settlement meant that they were always part of a larger system. The first Cuicatec Cañada irrigation of the high alluvium came about as a result of the direct intervention of the macrosystem. When the Zapotec withdrew, the Cañada towns may have been somewhat more self-sufficient, but as population began to grow in the Late Classic and Postclassic, it tied the new Cañada irrigation settlements to the highland settlements bordering them.

The same process of increasing population would be more difficult for the highland settlements to deal with. They could not turn heavily to irrigation, since, as the *relaciones* informed us, their major water resources were too entrenched to be widely deployed. Flat land was at a premium, and the construction of terraces was even more labor intensive than the work of the irrigation in the Cañada. Finally, in the higher fields only one crop a year would have been possible, due to frost.

As a result, the link between the lowland Cañada towns and the highlands was necessary from their inception. First, the highland towns had to depend on the lowland Cañada towns for food when the highland rainfall crops failed. Highland people, as we have seen from the historical sources, traded for lowland food, rented lands in the lowlands, or worked as wage labor in the lowland irrigation systems, to get by. Highland towns exchanged highland crops and products, like wood, for the lowland surpluses. They also specialized in weaving of cotton and making pottery. These specialties have no necessary link to the constraints of the highlands. However, they can be seen as an adaptation to the need to find some form of investment of labor that ultimately could be exchanged for the food that could not be produced by the highland fields. Finally, in the extreme cases when the Cañada towns refused to supply the food that was necessary for the existence of the highlands, the result might be war by the highlands to force the Cañada towns to come through. This was the cause of one war of the people of Almoloyas on the lowland towns tributary to them.

Consequently, the lowland towns always produced for the larger region. This, in part, was because the irrigation systems of the Cañada were always adapted for region-wide trade, not only of staples, but of specialties such as fruit. On the other hand, the lowland communities depended on highland products obtained through symbiotic exchanges, which included, among other things, specialized products like woven cotton. Thus the Cañada towns from their inception, were always part of a larger ecosystem, such as that which I defined in the historical analysis.

One argument that the Cuicatec towns were not self-sufficient agricultural communities is the evidence of stratification in the Cuicatec towns. The system of *caciques*, priests, nobles, commoners, and slaves is quite extensively described in the *relaciones*, particularly that of Atlatlauca. Archaeologically, in Cuicatlán there are the remains of considerable ceremonial and civic architecture, like the large mound and plaza, and *"juego de pelota,"* on the meseta of Iglesia Vieja, and the complex above the present town at Ojo del Agua. Even the small site associated with the terraces of El Despoblado had large architectonic features on it. These are the kinds of evidence that are traditionally interpreted as indicating a stratified society and which seem to indicate both civic and religious specialization.

One is tempted to tie this evidence of a hierarchical, stratified system to the irrigation system, as Wittfogel argues. The analysis of the resurgence of hierarchical systems in the Cañada towns based on historical sources suggests there are other factors that might play a part in this hierarchy.

Eva Hunt (1972) points out that the origin for the power of the most influential men in Cuicatlán was not necessarily land ownership, but rather the control of the highland-lowland trade. Now, as we have seen, from the beginning of the Cañada settlements, they were part of a larger ecosystem, which depended for its functioning on the flow of surplus agricultural produce from the Cañada irrigation systems to the other, highland parts of the systems, in exchange for other highland products. The elite would then be those who channeled the flow of these products, almost by definition.

From the historical analysis we argued that the hierarchical kind of system seemed to be related to the contact with the Mesoamerica-wide macrosystem. This kind of contact, when it existed, seemed to affect the Cañada systems more than the highland systems. However, the role of the larger, Mesoamerica-wide polities on the formation of the Cañada settlements is more problematical.

Susan Lees (1973) has suggested that there is a point at which the state does (at least in modern cases) intervene in local systems and produce changes that are irreversible, which in turn lead to greater and greater participation of the state in the local system. She recognizes that there are a number of local responses that seem to meet the needs of small irrigation systems that cropped up in Mesoamerica. At the same time she argues that there are situations when the larger state can take over control of local water resources because it is more efficient. One such situation is when more than one community shares the same stream and cannot effectively work out the problems caused by this conflict.

> Internal relations in communities are highly influenced (though not completely structured) by ties of kinship, friendship, and so forth, and these relations are generally sufficient to maintain the peace and reduce disputes. Such relations do not, however, generally obtain on any scale in intercommunity relations. It is one of the functions of the state to keep the peace between its component communities. It is particularly in its capacity as peace keeper that the state is effective in allocation: by keeping the peace and ensuring equitable distribution, the state enables the society to exploit water resources more effectively than it otherwise could. [Lees 1973:133]

Another such situation is when,

> (4) The use of water resources for canal irrigation proves a context in which the following social responses are likely to increase effectiveness in production: (a) mass or specialized labor on a large scale; (b) changes in community organization through state-level provision of services and facilities; (c) dependence of communities upon higher levels (or organic integration); (d) hierarchical controls of allocation and access; and (e) assertion of authority by higher or state-level organization. In other words, it is a context in which selection for those responses is likely to take place. [Lees 1973:133]

Specifically, she seems to have in mind two situations. First, she argues that the state tends to intervene in cases where two communities share the same water resource. In this case, the individual community is unable to cope with the conflict between the upstream and the downstream community, and the state has to step in. This is not necessary, but it is the most efficient way.

Her second argument is that there are certain larger agricultural works, specifically large dams, canals, etc., which demand the organization of the state to direct the workers to build them. Her examples are various modern irrigation projects, some of which have worked and others which have not. For the Cañada area, the Matamba dam and canal, which seem to have no precedents in the Cuicatec area in scale of conception and construction, represent this kind of intrusion of the state to introduce large-scale works which would be beyond the power of the individual communities. However, many of the conditions central to this argument simply did not apply to the Cañada area.

In the first place, it seems that in the kind of system of irrigation that obtained in all the Cuicatec systems, based on the Ríos Chiquitos, there simply were no cases of downstream communities where the upstream communities could control the water sources. Either the river source was of a size that it was too large (the Río Grande) or too small (the Ríos Chiquitos) and were only controlled by the downstream communities in the Cañada (for other reasons the upstream towns could not use up the waters, since the water was too entrenched and much of it came from the aquifers in the sandstone, which the upper towns could not control).

In the second place, there were no good places where large-scale mobilization of labor and of specialized skilled technicians could really improve the local irrigation scene. The Matamba canal, the local example of the kind of phenomenon that Lees is talking about, is illustrative in this way. While the present dam is large and impressive, the earlier technology managed to get water to the same area less dramatically. The present dam is already silted up, even though the canal is not completed. It seems, therefore, that at least in this area, this attempt of the national system to dominate the systems will be as ineffective as some of the failures which Lees describes for the Valley of Oaxaca.

The Cañada irrigation systems took their form from the direct intervention of the Preclassic-period Zapotec. The later contact with the expansionistic Aztec empire, as well as with earlier polities, was inevitable. We can see that the effects of this contact would have reinforced the tendency of the Cuicatec system to be hierarchical.

The contact was inevitable because the Cañada was a major north-south corridor. Active expansionistic armies would have to come down through the Cañada, and they did, as we saw from the historical sources. The small Cañada settlements would have been utterly incapable of successfully opposing these armies, which is probably why the conquest of this area attracted little notice in the Aztec histories. Before the armies actually came to the area, trade and other interaction of the major centers would have also passed through this area. When the Cañada settlements came into being, they would have been the first Cuicatec towns exposed to new influences, just as in historical times we saw that the Cañada towns were affected first and more strongly than the other towns by influences and innovations coming in from the Mesoamerica-wide system of the nation.

The domination of the Zapotec and Aztec systems would have tended to reinforce the class system in the Cañada towns. As we saw from the historical analysis, this would affect the Cañada towns more than the highland towns.

First, the Zapotec and Aztec made demands on the system, since the system would have to produce products for tribute beyond the needs of the Cuicatec ecosystem itself. These demands meant further intensification of the irrigation system and further strains on the highland agricultural systems. The *relaciones* specifically mention the need to produce tribute as the reason why highland people either worked as wage laborers in the lowlands or rented land in the lowlands.

The connection of the Cuicatec ecosystem to the larger system also may have widened the area to which Cuicatec Cañada products were traded. This may include the Cañada towns' trade in fruit all the way to the Valley of Oaxaca and the fruit they sent to the Aztec garrisons in Coixtlahuaca, in the Mixteca Alta. This long-distance trade would have been helped in part by the Pax Azteca. The highland production of woven cotton products could also be traded through the entire Mesoamerican network rather than remain only in the Cuicatec region. Both the tribute and the wider trade system would flow through the Cañada towns, because they were on the communication routes. Because the Cañada towns would be an amassing point of highland produce and the point through which the Zapotec and Aztec tribute was collected, through the local authorities, stratification would have been further reinforced there.

As we have observed, the domination of the Cañada towns in the Cuicatec ecosystem can be traced to irrigation, because the strength of the Cañada settlements is based directly on the surpluses that it could produce by intensive irrigation agriculture. This production tied it to the ecosystem and was its primary basis for trade beyond the Cuicatec ecosystem. However, irrigation does not seem to have been the "cause" of the rise of class structure, nor the basis for the power of the ruling elite. Rather than being the people who produced the most surplus, the role of the elite in the society seems to have been to channel the surpluses, both the interchanges from the highlands and lowlands and from the smaller system (the Cuicatec ecosystem) to the larger system (the Zapotec and Aztec empires). Power flowed, not from the irrigation canal directly, but rather from manipulation of trade and tribute, just as it does today.

Appendix 1

Archaeological Fieldwork

The archaeological fieldwork that I did in the Cuicatec region was primarily surface survey. This survey was most intensive in the vicinity of Cuicatlán. There was also a ten-day survey of other parts of the region. Surface collections were dated by comparison with known ceramic types from neighboring regions. In addition, two small excavations were undertaken to resolve questions that could not be answered from surface survey.

In the intensive survey around Cuicatlán the sites were located on aerial photographs and on large-scale (1:10,000) maps supplied by the Comisión de Papaloapan. It was hoped originally that the aerial photos could help locate the remains of the old irrigation systems. However, a combination of circumstances limited the usefulness of the aerial photos. The pre-Hispanic irrigation remains were preserved and could be distinguished only where they were higher than the present irrigated fields. However, the natural vegetation above the irrigated land is xerophytic scrub forest. This kind of vegetation has too coarse a texture to form crop marks reflecting features such as canals.

Numbered sites in general were simply collection sites. Collection sites were places where I made a collection of surface remains. In some cases this was because the remains were large and extensive. In other cases this was because the remains or the location of the site seemed atypical, and more information was desired about the site. Some sites were mapped, but no collections were made from them, in which case they usually were not given a number. A simple examination of the surface material in the field established the date of the site, and there was no need for closer analysis.

Some sites with large architectonic remains were mapped, in order to capture their size and form. Other sites were simply sketched in the notes, located on the map, and a collection made. The canal remains were located on maps, but no collections could be taken from them. They were, however, carefully drawn and photographed (See Figs. 15, 16, 17, 18, 25, 26, 27).

Two excavations were undertaken, in Site 5 and Site 16. The first excavation, at Site 5, was beside the pre-Hispanic canal. It was undertaken to establish the location and date of the canal. The second excavation, Site 16, attempted to establish a longer chronological sequence, as well as to recover samples for flotation. Flotation remains that were recovered are being analyzed by Judy Smith, of the University of Michigan. She has given me a preliminary report on the samples I have submitted to her.

The sites were dated mostly by the ceramics recovered from them. The ceramics were dated by comparison with types identified and fixed in time from archaeological research in the neighboring regions of Tehuacán, the Valley of Oaxaca, and the Mixteca Alta. Where sherds could not be related to types from these regions, they could not be dated. The ceramic artifacts are described in Hopkins 1973.

Survey of Cuicatlán, Oaxaca

Only sites from which collections were taken were given a number. I will first list and describe the numbered sites and then mention the other sites that were not given a number but were visited and noted. Figures 7 and 8 show the location of all these sites.

Site 1: Northern extreme of Valley of Cuicatlán, on small spur above the railroad tracks. At least two mounds and pottery concentrations.

Site 2: East of Site 1, on slope above irrigated fields. Pottery concentration.

Site 3: Northeast end of the Iglesia Vieja meseta. Mounds, house terraces, sacked tombs, pottery concentrations.

Site 4: East of Site 3, on the Iglesia Vieja meseta. Continuous foundations, pottery concentrations.

Site 5: House on small spur between the Iglesia Vieja meseta and the Río Chiquito. The neck of this spur was cut by the pre-Hispanic canal. One of the two excavations was made on this site.

Site 6: Sherd concentrations and house terraces on spur on southern side of the Río Chiquito, rising towards Site 10.

Site 7: Mound on one side of trail to La Sabana just before it crosses the Iglesia Vieja meseta.

Site 8: This site represented an attempt to collect ceramics from fields below Site 7. There were very few potsherds in the field.

Site 9: Crop mark in field identified from aerial photo. Collection from area of the crop mark revealed modern material, including roof tiles, which seem to indicate that a post-Hispanic structure had caused the mark.

Site 10: Large site that parallels the trail to Reyes Papalo atop the ridge between the cliff above Cuicatlán and the Río Chiquito.

Site 11: Site called "El Mirador," on Valerio Trujano side of the Río Grande, on first bank above lowest fields. Pottery concentration, Monte Albán I pottery.

Site 12: Small hill at the end of the Iglesia Vieja meseta, near airfield. Mounds, terraces, pottery concentrations. The site has been looted.

Site 13: On small hill behind Cuicatlán cemetery. Here a small L-shaped excavation had been put in, apparently by a professional archaeologist. The excavation was about 2.5 m long by 1.0 m wide, and about 1.0 m deep.

Site 14: Further up the hill behind the cemetery from Site 14. Plaster floors, foundations, pottery.

Site 15: On hill just across bridge over the Río Chiquito from Cuicatlán. House mound, foundations, pottery concentrations.

Site 16: Above cliff, on talus slope above the chapel at La Carbonera, south of Cuicatlán. Excavation made here (see Figs. 11, 14).

Site 17: Series of terraces across the barranca and south of Site 16, on next spur. One washed-out tomb, pottery concentrations.

Site 18: Across the barranca, on talus slope south of Site 17. Foundations, some looted tombs about 1.00 m by .60 m in dimensions. Pottery concentrations (see Fig. 12).

Site 19: Rock shelter in cliff south of Site 18. Here the bottom of a large pot was eroding out of the earth in the shelter. A flotation sample from this pot produced one small piece of pine and some modern grass seeds.

Site 20: Small sherd concentration on an eminence in the corridor past the Rinconada. Probably represented one small house.

Site 21: Ojo de Agua—on talus slope directly above Cuicatlán. Trail to Reyes Papalo winds through this site. Large terraces, a well, large mound, and part of a stairway (see Figs. 9, 10).

Site 22: On talus slope below Site 16. Large mound on a platform. The platform is 40 m by 85 m, oriented with the long axis north-south. At the northeast corner of the platform is a notch 12 m deep by 28 m wide.

Site 23: Terraces, foundations, on talus spur below Sites 17 and 18. Heavy sherd concentration.

Site 24: Another large platform, terraces, and foundations, on talus slope south of Site 23 and below Sites 17 and 18.

Site 25: House foundation being dug by resident of Cuicatlán just below the site of Ojo de Agua (Site 21), which turned up flat rectangular clay plaques with petate impressions, and gray pottery tubes (see Hopkins 1973:15-17).

Site 31: Material found by workers excavating the Matamba canal, at Km 3.200. Sherds, but no foundations, were reported.

All of the collections from the above sites, with the exception of Site 11, were almost entirely Postclassic material. In addition to these remains, a check of the low hills in Valerio Trujano, across the river from Cuicatlán, revealed that they were also covered with foundations and pottery concentrations from the Postclassic period.

Figure 8 shows that almost all of the sites were located on eminences above irrigable land. This is not the result of inadequate survey. Several attempts were made to locate sites both in present irrigated fields, and in land that might have been irrigated by earlier, higher canals, especially in the area to the north of the meseta of Iglesia Vieja. First, I wanted to make sure that I was not missing occupation in the flat irrigated land. Secondly, I hoped to locate earlier sites, especially from the Preclassic, since the one site from this period was located down on the first terrace above the river (Site 11).

However, these efforts were unsuccessful. The fields above the floodplain of the Río Grande had a very thin scatter of small, broken-up sherds mixed through them. Nothing like a concentration of artifacts was located. These small scattered sherds probably represented remains of artifacts carried down to the fields by agricultural workers. The lack of concentrations indicate there were probably no permanent structures among the fields. Four hundred years of cultivation since has scattered the few remains through the fields.

On aerial photos a crop mark that seemed to indicate a potential site in irrigable land was investigated (Site 9). The remains recovered from it were mostly modern and included roof tiles. The crop mark was probably the result of a post-Hispanic structure that has since disappeared.

Down in the floodplain proper of the Río Grande, examination of the fields did not even reveal the thin scatter of sherds found up above the floodplain on the fields. Soil in the Río Grande floodplain was noticeably more sandy and less loamy than that above the first river terrace. The Río Grande has reworked this soil enough during 400 years of floods that all remains have been destroyed.

One of the major aims of this project was to recover remains of agricultural systems, in particular of old canals. In this I was successful. Remains of pre-Hispanic canals were located on the Río Chiquito of Cuicatlán, the Cacahuatal, between Cuicatlán and Quiotepec, and across the river from Cuicatlán, near the mouth of the Apoala, in Valerio Trujano. This last site had three levels of canals. The highest was the pre-Hispanic canal of rock and mortar construction. The next highest was a post-Hispanic brick aqueduct built into a bridge across the Apoala that is now in ruins. Finally, the modern canal is considerably lower than either of the earlier constructions.

On the northern side of the Río Chiquito of Cuicatlán, about a kilometer upstream from the present *toma de agua* on that side of the Río Chiquito are fragments of another pre-Hispanic masonry canal. These fragments were built into the hollow of the cliff that borders the bed of the Río Chiquito. At the point that the *toma de agua* for this canal must have existed, the canyon of the Río Chiquito was entirely closed in, and there is no floodplain between the walls of the canyon and the stream bed, even in the dry season (Figs. 15, 16).

Two of these fragments had remnants of the canal bed. By leveling with a hand level, the average drop of the canal bed between these two fragments could be established. This drop turned out to be 1.1%. Since along the same distance the drop of the river was about 3.5%, the downstream fragments of the canal were progressively higher on the slope of the canyon above the Río Chiquito (Fig. 16). Eventually this canal cut the neck of the small spur on which Site 5 had been located. Subsequent excavation established that this site was Postclassic.

Another well-preserved canal fragment like those on the Río Chiquito of Cuicatlán was found on the Cacahuatal between Cuicatlán and Quiotepec (Fig. 25).

Other Cuicatec Sites

A part of the study was a 10-day survey of the whole Cañada. The purpose of this trip was to see to what extent the data from Cuicatlán were characteristic of the entire Cañada. During this trip, all the important towns of the Cañada were visited, as well as the highland towns of Nacaltepec and Cotahuixtla, and samples of surface pottery were collected. Particular emphasis was placed on looking for early sites. I was aided enormously on this trip by the archaeological inspector for the I.N.A.H., Rafael Vásquez Cruz, who, through 30 years of diligent work, has acquired an intimate knowledge of the Cuicatec area. He was able to take me directly to most of the areas of interest, saving me an enormous amount of time. In addition, his acquaintance with the routes, places to stay, places to eat, and local officials was essential to the success of the trip.

On this survey, sites in Nacaltepec, Atlatlauca, Cotahuixtla, Dominguillo, and El Chilar were visited (Fig. 7). This circuit was made partly on foot and partly in trucks and the public bus that goes down from Nacaltepec to Cuicatlán once a day. Sites were visited in highlands and lowlands, as well as a few in the zone in between.

The survey began in Nacaltepec, in the highlands above Atlatlauca. Buried check dams were found in the arroyos there. They are not kept up today and have been cut through in places by the arroyo. There is also a large Postclassic hilltop site, with one well-preserved foundation of limestone of a large building.

El Sabino

On the road from Nacaltepec to Atlatlauca, before the dropoff at the Peña de Ejutla, is a small site called El Sabino. This site consists of a few mounds and sacked tombs. The pottery on the surface included Postclassic gray vessels, one piece of *fondo sellado*, and several effigy feet.

La Unión (Site 28)

When the road from Nacaltepec to La Unión descends to the floor of the canyon of the Río de las Vueltas, it goes through a large flat area of irrigated fields. Just as it reaches the Río de las Vueltas, across and upstream from Atlatlauca, the road makes a bend and follows the river downstream. At the point where the road turns are two large groups of mounds. These are on the flat, on the first terrace above the river. The sherds collected from this site were like those described by Spencer (1982) for the Lomas phase. The site is interesting because of its size and location, where the road from Telixtlahuaca enters the Cañada.

Atlatlauca

The archaeological inspector, Rafael Vásquez Cruz, reported a large site high on top of a steep hill behind the present town. Later examination of collections made by him from this site revealed Postclassic material, interestingly with a large amount of "Mixtec polychrome," which was rare in all the other collections I made from the Cuicatec region. On my visit I concentrated on ascending the Río Chiquito of Atlatlauca to look for remains of old canals. Despite a day ascending both branches of the Río Chiquito at Atlatlauca, no such remains were found.

Cotahuixtla

On the hills behind Cotahuixtla, in oak-pine forest, were the remains of an abandoned terrace system. There was no evidence as to its date.

El Despoblado and the Río de las Trancas Canals (Site 32)

One of the tributaries of the Río de las Vueltas starts just behind the town of Cotahuixtla, and joins the Río de las Vueltas just above Dominguillo. This tributary is called the Río de las Trancas or the Carrizal. On both sides of this river are remains of masonry canals. These canals are partially consolidated with carbonates from the waters, which flow through limestone bedrock at this point. One of the canals, on the northern side of the river, clings to the sheer cliff face (Fig. 27b), and can be traced down to a series of terraces, stuck on the side of a steep ridge. There are 26 of these terraces (Fig. 28) and one small house mound associated with them.

On the ridge above these terraces is the site of El Despoblado. This site runs along the ridge and contains continuous occupation remains, as well as at least one large, well-preserved structure (Fig. 29). This was associated with Postclassic material.

La Coyotera (Site 29)

North of Dominguillo was a small site covering one of the small hills that rose from the land along the canyon of the Río de las Vueltas. This site had some mounds and terraces on its surface, associated with Postclassic sherds, as well as a few sherds of earlier periods (see Spencer 1982; Redmond 1983). Behind this hill, and connecting it with another small hill, is a small abandoned stone construction, about 155 m long, 2 or 3 m high, and 3 or 4 m wide. This appears to have been a small aqueduct to keep an earlier canal at a high level across the saddle between these two hills.

Ojo de Pajarito

Another small site near Dominguillo is next to a small spring. This spring only has enough water for irrigation every five or six years. This year, since there had been record-breaking rains the preceding rainy season, a small irrigation ditch led down to just behind La Coyotera. The site is a small mound and a platform, associated with Postclassic pottery.

Santa Cruz

This is a small hamlet on the Mixtec slope behind Dominguillo. It is at the head of the Barranca del Aguacate, above where the drinking water for Dominguillo is taken off. The hamlet is perched on a limestone cliff just above the contact between the limestone and the reddish conglomerates that underlay it. There is a small irrigation system with terraces and tanks, fed by a spring in a cave above the hamlet. Above the present town the archaeological inspector reports a large site, which he had been requested by his superiors not to show to people.

In the barranca below the spring are abandoned

check dams and terraces. On the slope of the irrigated fields, paralleling the present canals, are remains of older, mineralized canals (Fig. 26), reminiscent of those reported from the Valley of Tehuacán and from Hierve el Agua (Neely and Woodbury 1972; Neely 1967).

Dominguillo

An attempt to find an abandoned pre-Hispanic canal above Dominguillo failed. The present canal is taken off the Río de las Vueltas and is of masonry, built into the cliff face much like the pre-Hispanic fragments at Cuicatlán. The old *toma de agua* may not have been any higher than the present one, because the canyon of the Río de las Vueltas closes in just above Dominguillo. The river fills this canyon from wall to wall every rainy season and threatens any construction within reach of its floods.

Directly across the river from Dominguillo is an adobe platform and mound, on a point surrounded on three sides by the river. This Postclassic site is noteworthy for its location.

Excavations

Site 5

The route marked by the fragments of the canal on the Río Chiquito of Cuicatlán led to a point where it apparently cut the neck of a small spur that had what seemed to be an artificial mound on it. This offered a chance to date the canal. An excavation was made in this site, Site 5, for two purposes. First, I wanted to establish that the canal did in fact cut the spur where I thought it did. Secondly, I wanted to establish the date and the nature of the site on the end of the spur, in order to date the canal.

In order to verify that the canal did cut the spur, I put in a trench 11 m long by 1 m wide, and a little over 2.5 m deep at the deepest, across the expected path of the canal (Figs. 19, 20). Figure 18 shows the profile from this trench. The bedrock is loosely consolidated conglomerate, which at first was mistaken for pavements of rock. At this point it was recognized that the apparent terrace walls that formed the edge of the mound on the end of the spur really represented where horizontal layers of conglomerate with large rocks intersected the surface.

An excavation was made on the site on the end of the spur cut off by the canal. Barely beneath the surface of the ground this excavation uncovered a plaster house floor (Fig. 21). This plaster floor had a low step running across it about 7 cm high, so that one part of the floor was higher than the other (Fig. 23). On one side of the floor one course of rocks about 3 m long remained of a wall. Five meters across from this wall was another, parallel wall some two or three courses high which formed the other side of the house (Fig. 22). In line with the shorter wall segment was a series of postholes, some with rocks in them (Fig. 21). Outside the more complete wall was a shallow pit that seemed to be a fireplace, filled with charcoal and a number of sherds. The flotation sample from this fireplace contained identifiable seeds, corn, nutshells, as well as charcoal from pine and other woods (Judy Smith, personal communication). This feature was on the side of the house that was closest to the canal, and probably was the place where a kitchen lean-to had stood. On the other side of the house were the sherds of what turned out to be half of a large, coarse olla. The other half had undoubtedly been exposed to the surface and destroyed. It seemed to have stood just outside the house. On the plaster floor was a nearly complete "patojo," or shoe-shaped vessel. Because the house was so close to the surface of the ground, further reconstruction was impossible. Sherds from the fill associated with the floor and the walls of the house established that it was Postclassic. Its walls and floor were oriented with the canal.

A profile was cut through the plaster floor of the house, to establish if there were more occupation layers. This revealed two tombs, which shared a common wall (Fig. 24). These had also eroded considerably, so that the walls of the tomb only stood one course high. Some of the walls of the tomb had been entirely eroded away. Tomb 1 contained one individual in very poor condition and a number of gray vessels, all of which were smashed. The second tomb had the remains of two individuals in it, also in a very powdery condition, and four flat-bottomed gray bowls. These tombs were oriented with the floor above them, and with the canal. The pottery in the tombs was all of Postclassic gray types.

Site 16

Site 16 is a large occupation site with some large mounds (Fig. 11). It is high on the talus slope above the La Carbonera chapel in Cuicatlán. I began an excavation there with two goals in mind. The first was to obtain more samples from flotation, to attempt to establish evidence of agricultural products, as well as of products from the highlands. Secondly, this site had yielded some of the possible Classic-period "strays," and I hoped that I might get a chronological sequence, reaching earlier periods at lower levels of the site.

In order to investigate this area I opened up a trench 8 m by 1 m on a north-south line. Later I expanded the southern 3 m of this trench 1 m to the east. Finally I dug down in the two southernmost squares of this excavation.

The first trench landed squarely on top of a step in a plaster floor, like that encountered in the excavation on Site 5 (Fig. 14a). This step was also about 7 cm high. The higher part of the stepped floor was underlain by a pavement of rocks, unlike that at Site 5.

In the first level there was an alignment of rocks that seemed to represent something like a channel or a gutter. This feature was aligned with the step in the plaster floor. On each side of it were two nearly complete vessels. One of these was a nearly complete polychrome tripod olla, which along with only ten other sherds, represented the total amount of polychrome recovered from the entire region.

Cutting down through the first plaster floor and step, I found that the rocks that delineated the edge of the step followed the line of a lower floor. Beneath these rocks (Fig. 14b, c) the lower plaster floor outlined a corner. Beneath this second floor, but not as precisely oriented with it, was a layer of gray ashy material with lots of animal bone and sherds, as well as rock. This feature could not be interpreted because of the small area exposed.

Beneath this was a terre pisé wall of adobe-like reddish sandy dirt (Fig. 14d). This wall had a plaster floor on each side of it. These floors and this wall were not oriented with the upper two floors. Limitations of time and money forced the termination of the excavation at this level.

Flotation material was obtained from the ashy layer beneath the second plaster floor. Although two or three building levels were represented in this excavation, sherds from beneath the lowest plaster floors were late Postclassic Huitzo Red on Cream. Consequently, no long chronological sequence was established. The architectural features are also hard to interpret because of the limited area exposed. The unique plaster step that occurs on both sites is interesting.

Artifacts

Most of the material collected consisted of ceramics. This was in part because the purpose of the collections was to date sites, and ceramics serve this purpose best. The ceramic artifacts are described in Hopkins 1973. There were in all only 69 non-ceramic artifacts out of a total of 5086 artifacts. Twenty-eight of these non-ceramic artifacts came from the excavations. This proportion is probably representative. In the excavations, from which more than half of the ceramics came, everything was kept. In the surface collections lithic artifacts were collected whenever they occurred because they were so rare. There were very few lithic objects on the sites.

Most of the material was obsidian. There were 52 obsidian artifacts. There were 7 quartz artifacts and only 2 of chert. In addition there were 6 flat, circular slate beads, with biconical holes drilled in them. They ranged from 1.5 to 3 cm in diameter. No two were from the same site. Most were of gray slate, although one was bluish and another of green stone.

One broken cylindrical bone bead, about 1 cm in diameter, was recovered from the excavation of Site 16. A fragment of a metate was found in Site 10, on the top of the cliff behind Cuicatlán.

The two flint pieces were flakes. Five of the quartz pieces were flakes or chunks. However, two were carefully worked points of projectiles or knives.

Most of the obsidian was gray. Some of this gray obsidian was almost transparent. The three cores found were gray. Some of the obsidian was green, including a few pieces with a golden surface like that characteristic of late assemblages from the Valley of Mexico. There were also two pieces of opaque black obsidian.

Of the 52 pieces of obsidian, 32 were blades that showed little or no edge wear. There were 4 flakes of obsidian that also showed no more than edge wear. Seven blades had been retouched. Three were retouched from the ventral side, and 3 from the dorsal side. One had alternate retouch. Six scrapers were made on blades, and 1 denticulate-looking tool was made on an obsidian flake.

All of the material collected, except the animal bones and the flotation samples, which are still being analyzed, were deposited in the museum in the city of Oaxaca.

Bibliography

Adams, Robert McC.
1961   Changing Patterns of Territorial Organization in the Central Highlands of Chiapas, Mexico. *American Antiquity* 26:341-60.

Adams, Robert McC., and Hans J. Nissen
1972   *The Uruk Countryside: The Natural Setting of Urban Society*. Chicago: University of Chicago Press.

Ajofrín, Francisco de
1959   *Diario del Viaje que . . . Hizo a la América Septentrional en el Siglo XVIII el P. Fray Francisco Ajofrín, Capuchino*. Archivo Documental Español 13. Madrid: Academia de la Historia.

Altolaguirre y Duvale, Ángel de
1954   *Descubrimiento y Conquista de México*. México: Salvat Editores.

Armillas, Pedro
1949   Notas Sobre Sistemas de Cultivo en Mesoamérica: Cultivos de Riego y Humedad en la Cuenca del Río de las Balsas. *Anales del Instituto Nacional de Antropología e Historia* 3:85-113. México.

1971   Gardens on Swamps. *Science* 174:653-61.

Armillas, Pedro, and Robert West
1950   Las Chinampas de México. *Cuadernos Americanos* 2:165-82.

Attolini, José
1941   *Economía de la Cuenca del Papaloapan: Agricultura*. México: Instituto de Investigaciones Económicas.

1950   *Economía de la Cuenca del Papaloapan: Bosques, Fauna, Pesca, Ganadería, e Industria*. México: Instituto de Investigaciones Económicas.

Bandelier, Adolf F.
1884   Report of an Archaeological Tour in Mexico in 1881. *Papers of the Archaeological Institute of America* 2.

Barlow, Robert H.
1949   The Extent of the Empire of the Culhua-Mexica. *Ibero-Americana* 28.

Barrera, Tomás
1946   *Guía Geológica de Oaxaca*. México: Universidad Nacional Autónoma de México, Instituto de Geología.

Bateson, Gregory
1972   Culture Contact and Schismogenesis. In *Steps to an Ecology of Mind*, pp. 61-72. New York: Ballantine Books.

Belmar, Francisco
1901   *Breve Reseña Histórica y Geográfica del Estado de Oaxaca*. Oaxaca: Imprenta del Comercio.

1902   *El Cuicateco*. Oaxaca: Imprenta del Comercio.

Bernal, Ignacio
1966   Ruinas de Santo Domingo, Oaxaca. *Boletín del Instituto Nacional de Antropología e Historia* 24:8-12.

Boletín
1975   *Boletin*, no. 2, Mayo 1975. Centro Regional de Oaxaca, Instituto Nacional de Antropología e Historia, Oaxaca, Oaxaca, México.

Borah, Woodrow
1943   Silk Raising in Colonial Mexico. *Ibero-Americana* 20.

Borah, Woodrow, and Sherburne F. Cook
1958   Price Trends of Some Basic Commodities in Central Mexico, 1531-1570. *Ibero-Americana* 40.

1960   The Population of Central Mexico in 1548—An Analysis of the 'Suma de Visitas de Pueblos'. *Ibero-Americana* 43.

1963   The Aboriginal Population of Central Mexico on the Eve of the Spanish Conquest. *Ibero-Americana* 45.

Boserup, Ester
1965   *The Conditions of Agricultural Growth: The Economics of Agrarian Growth under Population Pressure*. Chicago: Aldine.

Bradomín, José María
1955   *Toponimia de Oaxaca*. México.

British Honduras Land Use Survey Team
1959   Land in British Honduras. Report of the British Honduras Land Use Survey Term. 2 Vols. D.H. Romney, ed. Colonial Research Publications 24. London: H.M. Stationery Office.

Burgoa, Francisco de
1934a *Geográfica Descripción . . . de esta Provincia de Antequera, Valle de Guaxaca*. [1674]. 2 vols. Publicaciones del Archivo General de la Nación 25-26. Mexico: Talleres Gráficos de la Nación.

1934b *Palestra Historial de Virtudes y Ejemplares Apostólicos . . .* [1670]. Publicaciones del Archivo General de la Nación 24. Mexico: Talleres Gráficos de la Nación.

Caso, Alfonso, Ignacio Bernal, and Jorge R. Acosta
1967   *La Cerámica de Monte Albán*. Memorias del Instituto Nacional de Antropología e Historia 13. México.

Cerda Silva, Roberto de la
1942   Los Cuicatecos. *Revista Mexicana de Sociología* 4(4).

Chavero, Alfredo, ed.
1892   Códice Porfirio Díaz. In *Antigüedades Mexicanas Publicadas por la Junta Colombina de México en el Cuarto Centenario del Descubrimiento de América*, Vol. 1:xi-xix and 21 plates in Atlas. México: Tipográfica de la Secretaría de Fomento.

Chevalier, François
1956   La Formación de los Grandes Latifundios en México (Tierra y Sociedad en los Siglos XVI y XVII). *Problemas Agrícolas e Industriales de México* 8:1-258. México.

Childe, V. Gordon
1950   The Urban Revolution. *Town Planning Review*

21:3-17.

Cline, Howard F.
1972 A Census of the Relaciones Geográficas of New Spain, 1579-1612. In *Handbook of Middle American Indians*, Robert Wauchope, gen. ed., Vol. 12: *Guide to Ethnohistorical Sources, Part One*, Howard F. Cline, vol. ed., pp. 324-69. Austin: University of Texas Press.

Codex Fernández Leal—see Peñafiel 1895.

Codex Porfirio Díaz—see Chavero 1892.

Coe, Michael D., and Kent V. Flannery
1964 Microenvironments and Mesoamerican Prehistory. *Science* 143:650-54.

Cook, Sherburne F.
1949 The Historical Demography and Ecology of the Teotilalpan. *Ibero-Americana* 33.
1968 The Population of the Mixtec Alta, 1520-1960. *Ibero-Americana* 50.

Cook, Sherburne F., and Woodrow Borah
1960 The Indian Population of Central Mexico—1531-1610. *Ibero-Americana* 44.

Cook, Sherburne F., and Lesley B. Simpson
1949 The Population of Central Mexico in the Sixteenth Century. *Ibero-Americana* 31.

Cortés, Hernán
1866 *Cartas y Relaciones de Hernan Cortés al Emperador Carlos V.* Colegida e ilustradas por Don Pascual de Gayangos. Paris: Imprenta Central de los Ferrocarriles, A. Chaix y Cia.
1963 *Cartas y Documentos.* Introducción de Mario Hernández Sánchez Barba. Mexico: Editorial Porrúa.

Cowgill, Ursula M.
1962 An Agricultural Study of the Southern Maya Lowlands. *American Anthropologist* 64:273-86.

Cuadros Sinópticos—see Martínez Gracida 1883.

De la Cerda Silva, Roberto—see Cerda Silva, Roberto de la.

Díaz del Castillo, Bernal
1967 *Historia Verdadera de la Conquista de la Nueva España.* Introducción y notas de Joaquín Ramírez Cabañas. 5th ed. México: Editorial Porrúa.

Dirección General de Estadística, México
1886, 1898, 1900, 1910, 1921, 1930, 1940, 1950,
1960 Censo General de la República Méxicana. México, D. F.

Donkin, R. A.
1970 Precolumbian Field Implements and Their Distribution in the Highlands of Middle and South America. *Anthropos* 65:505-29.

Drennan, Robert D.
1978 Excavations at Quachilco: A Report on the 1977 Season of the Palo Blanco Project in the Tehuacán Valley. *Museum of Anthropology, University of Michigan, Technical Reports* 7.

Drennan, Robert D., ed.
1977 The Palo Blanco Project: A Report on the 1975 and 1976 Seasons in the Tehuacán Valley. Andover: R. S. Peabody Foundation for Archaeology. Ann Arbor: University of Michigan Museum of Anthropology.
1979 Prehistoric Social, Political, and Economic Development of the Area of the Tehuacán Valley: Some Results of the Palo Blanco Project. *Museum of Anthropology, University of Michigan, Technical Reports* 11.

Durán, Fray Diego
1965 *Historia de las Indias de Nueva España.* 2 vols. México: Editora Nacional.

Esteva, Cayetano
1913 *Nociones Elementales de Geografía Histórica del Estado de Oaxaca.* Oaxaca: Tip. San Germán Hnos.

Flannery, Kent V.
1968a The Olmec and the Valley of Oaxaca: A Model for Interregional Interaction in Formative Times. In *Dumbarton Oaks Conference on the Olmec*, Elizabeth P. Benson, ed., pp. 79-110. Washington: Trustees for Harvard University.
1968b Archaeological Systems Theory and Early Mesoamerica. In *Anthropological Archeology in the Americas*, Betty J. Meggers, ed., pp. 67-87. Washington: Anthropological Society of Washington.
1972 The Cultural Evolution of Civilizations. *Annual Review of Ecology and Systematics* 3:399-426.

Flannery, Kent V., and Joyce Marcus
1976 Formative Oaxaca and the Zapotec Cosmos. *American Scientist* 64(4):374-83.

Flannery, Kent V., and Joyce Marcus, eds.
1983 *The Cloud People: Divergent Evolution of the Zapotec and Mixtec Civilizations.* New York: Academic Press.

Flannery, Kent V., Joyce Marcus, and Stephen A. Kowalewski
1981 The Preceramic and Formative of the Valley of Oaxaca. In *Supplement to the Handbook of Middle American Indians*, Victoria R. Bricker, gen. ed., Vol. 1: *Archeology*, Jeremy A. Sabloff, vol. ed., pp. 48-93. Austin: University of Texas Press.

Flannery, Kent V., Anne V. T. Kirkby, Michael J. Kirkby, and Aubrey Williams, Jr.
1967 Farming Systems and Political Growth in Ancient Oaxaca. *Science* 158:445-54.

Fosberg, R. R.
1945 Principal Economic Plants of Tropical America. In *Plants and Plant Science in Latin America*, Frans Verdoorn, ed., Vol. 16:18-35. Waltham, Mass.: Chronica Botanica.

Fried, Morton H.
1967 *The Evolution of Political Society.* New York: Random House.

Gay, José Antonio
1881 *Historia de Oaxaca.* 2 Vols. México.

Gerhard, Peter
1972 *A Guide to the Historical Geography of New Spain.* Cambridge Latin American Studies 14. David Joslin, Timothy Knog, Clifford T. Smith, and John Street, eds. Cambridge: Cambridge University Press.

Gibson, Charles
1964 *The Aztecs Under Spanish Rule.* Stanford: Stanford University Press.

Hamnett, Brian R.
1971 *Politics and Trade in Southern Mexico, 1750-1821.* Cambridge: Cambridge University Press.

Herrera y Tordesillas, Antonio de
1726 Historia de las Indias Occidentales. Madrid.
1730 Historia General de los Hechos de los Castellanos en las Islas; Tierra Firme del Mar Océano. 5 Vols. Madrid: Oficina Real de Nicolas Rodriguez Franco.

Holland, William R., and Robert J. Weitlaner
1960 Modern Cuicatec Use of Prehistoric Sacrificial Knives. *American Antiquity* 25:392-448.

Hopkins, Joseph W., III
1968a Prehispanic Agricultural Terraces in Mexico.

BIBLIOGRAPHY

            M.A. thesis, University of Chicago.
1968b   A Prehispanic Agricultural System in South Central Mexico. Ph.D. thesis proposal, University of Chicago.
1973   Ceramics of La Cañada, Oaxaca, Mexico. *Vanderbilt University Publications in Anthropology* 6.
1974   Irrigation and the Cuicatec Ecosystem: A Study of Agriculture and Civilization in North Central Oaxaca, Mexico. Ph.D. Dissertation, University of Chicago.
1983   The Tomellín Cañada and the Postclassic Cuicatec. In *The Cloud People: Divergent Evolution of the Zapotec and Mixtec Civilizations*, Kent V. Flannery and Joyce Marcus, eds., pp. 266-70. New York: Academic Press.

Hunt, Eva
1972   Irrigation and the Socio-political Organization of the Cuicatec Cacicazgos. In *The Prehistory of the Tehuacán Valley*, Vol. 4: *Chronology and Irrigation*, Frederick Johnson, ed., pp. 162-248. Austin: University of Texas Press.

Hunt, Eva, and Robert C. Hunt
1973   Irrigation, Conflict, and Politics: A Mexican Case. Unpublished manuscript.

Hunt, Robert C., and Eva Hunt
1973   Irrigation and Local Social Structure. Paper presented at the 1973 Meeting of the American Anthropological Association, New Orleans.
1976   Irrigation, Conflict, and Politics: A Mexican Case. In *Irrigation's Impact on Society*, Theodore Downing and McGuire Gibson, eds., pp. 129-57. Tucson: University of Arizona Press.

Iturribarría, Jorge Fernando
1955   *Oaxaca en la Historia*. México: Editorial Stylo.

Ixtlilxóchitl, Fernando de Alva
1965   *Historia Chichimeca; Obras Históricas de Don Fernando de Ixtlilxóchitl*, Vol. 2, Publicadas y anotadas por Alfredo Chavero. México: Editora Nacional.

Jenny, Hans
1958   Role of the Plant Factor in the Pedogenic Functions. *Ecology* 39:5-16.
1961   Derivation of State Factor Equations of Soils and Ecosystems. *Soil Science Society Proceedings, 1961*: 385-88.

Johnson, Frederick, and Richard S. MacNeish
1972   *The Prehistory of the Tehuacán Valley*, Vol. 4: *Chronometric Dating*. Austin: University of Texas Press.

Kirkby, Anne V.T.
1973   The Use of Land and Water Resources in the Past and Present Valley of Oaxaca, Mexico. Prehistory and Human Ecology of the Valley of Oaxaca, Kent V. Flannery, gen. ed., Vol. 1. *Memoirs of the Museum of Anthropology, University of Michigan* 5.

Leach, Edmund R.
1964   *Political Systems of Highland Burma*. Boston: Beacon Press.

Lees, Susan H.
1973   Sociopolitical Aspects of Canal Irrigation in the Valley of Oaxaca. Prehistory and Human Ecology of the Valley of Oaxaca, Kent V. Flannery, gen. ed., Vol. 2. *Memoirs of the Museum of Anthropology, University of Michigan* 6.

El Libro de las Tasaciones de Pueblos de la Nueva España, Siglo XVI
1952   *El Libro de las Tasaciones de Pueblos de la Nueva España, Siglo XVI*. México: Archivo General de la Nación.

MacNeish, Richard S.
1958   Preliminary Archaeological Investigations in the Sierra de Tamaulipas, Mexico. *Transactions of the American Philosophical Society*, N.S., 48 (Pt. 6).
1967   An Interdisciplinary Approach to an Archaeological Problem. In *The Prehistory of the Tehuacán Valley*, Vol. 1: *Environment and Subsistence*, Douglas W. Byers, ed., pp. 14-24. Austin: University of Texas Press.

MacNeish, Richard S., Frederick A. Peterson, and Kent V. Flannery
1970   *The Prehistory of the Tehuacán Valley*, Vol. 3: *Ceramics*. Austin: University of Texas Press.

MacNeish, Richard S., Melvin L. Fowler, Ángel García Cook, Frederick A. Peterson, Antoinette Nelken-Terner, and James A. Neely
1972   *The Prehistory of the Tehuacán Valley*, Vol. 5: *Excavations and Reconnaissance*. Austin: University of Texas Press.

Marquina, Ignacio
1951   Arquitectura Prehispánica. *Memorias del Instituto Nacional de Antropología e Historia* 1. México.

Martínez, Maximino
1948   Algunas Observaciones Relativas a la Flora de Cuicatlán, Oaxaca. *Anales del Instituto de Biología de México* 19(2):365-91.

Martínez Gracida, Manuel
1883   *Colección de Cuadros Sinópticos de los Pueblos, Haciendas y Ranchos del Estado Libre y Soberano de Oaxaca*. Anexo No. 50 a la Memoria Administrativo presentado al H. Congreso del Mismo el 17 de Septiembre de 1883. Oaxaca: Imprenta del Estado.

Meade, Joaquín
1939   Exploraciones en la Huasteca Potosina. *Proceedings of the 27th International Congress of Americanists*, Mexico City, Vol. 2:12-24.

Metcalfe, Grace
1946a   Índice de la Palestra Historial. *Boletín del Archivo General de la Nación* 17(4). México.
1946b   Índice a la Geográfica Descripción. *Boletín del Archivo General de la Nación* 17(4). México.

México, Dirección General de Estadística—see Dirección de Estadística.

Millon, René
1957   Irrigation Systems in the Valley of Teotihuacán. *American Antiquity* 22:160-66.
1962   Variations in Social Responses to the Practice of Irrigation Agriculture. In: Civilization in Arid Lands, Richard B. Woodbury, ed., pp. 56-88. *University of Utah Anthropological Papers* 62.
1967   Teotihuacán. *Scientific American* 216:38-48.

Miranda, F.
1948   Datos sobre la Vegetación en la Cuenca Alta del Papaloapan. *Anales del Instituto de Biología de México* 19(2):333-64.

Miranda, José
1952   *El Tributo Indígena en la Nueva España durante el Siglo XVI*. México: Fondo de Cultura Económica.

Muñoz Lumbier, Manuel, and Alberto Quintanar
1935   *Las Zonas Inclementes de Oaxaca*. México: Talleres Gráficos de la Nación.

Neely, James A.
1967   Organización Hidráulica y Sistemas de Irrigación Prehistóricos en el Valle de Oaxaca. *Boletín del*

Instituto Nacional de Antropología e Historia 27.
Oficina de Registro Civil, Cuicatlán
   1882   Censo de Habitantes hasta el fin del Año 1882. (D. Fidel Jiménez.) Xerox of manuscript loaned by Eva Hunt.
Palerm, Ángel
   1954   La Distribución del Regadío en el Área Central de Mesoamérica. *Ciencias Sociales* 5:2-15. Washington, D.C.
   1955   The Agricultural Basis of Urban Civilization in Mesoamerica. In *Irrigation Civilizations: A Comparative Study*, Julian Steward, ed., pp. 28-42. Social Science Monographs 1. Washington, D.C.: Pan American Union.
   1961   Agricultura de Riego en el Viejo Señorío del Acolhuacan. *Pan American Union, Revista Interamericana de Ciencias Sociales*, segunda época, 1(2). Washington, D.C.
   1967   Agricultural Systems and Food Patterns. In *Handbook of Middle American Indians*, Robert Wauchope, gen. ed., Vol. 6: *Social Anthropology*, Manning Nash, vol. ed., pp. 26-52. Austin: University of Texas Press.
Palerm, Ángel, and Eric Wolf
   1957   *Ecological Potential and Cultural Development in Mesoamerica*. Studies in Human Ecology, Social Science Monographs 3. Washington, D.C.: Pan American Union.
Pareyón Moreno, Eduardo
   1960   Exploraciones Arqueológicas en Ciudad Vieja de Quiotepec, Oaxaca. *Revista Mexicana de Estudios Antropológicos* 16:97-104.
Parsons, Jeffrey R.
   1968   Teotihuacán, Mexico, and Its Impact on Regional Demography. *Science* 163:872-77.
   1971   Prehistoric Settlement Patterns in the Texcoco Region, Mexico. *Memoirs of the Museum of Anthropology, University of Michigan* 3.
Paso y Troncoso, Francisco del, ed.
   1905   *Papeles de Nueva España: Segunda Serie, Geografía y Estadística*. 7 Vols. Madrid: Est. Tipográfico "Sucesores de Rivadeneyra."
   1946   *G^{al} indica d'todas las Jurrisdicion^s d'esta N^a Esp.^{a} Frutos y Obisp.^{dos} Tributos y Tributarios 1784*. Biblioteca Aportación Histórica. México: Editorial Vargas Rea.
   n.d.   Relaciones Geográficas del Siglo XVIII (1778) Cuicatlán. Pts. 1 and 2, Vol. 99, Papalo Vol. 100. Copia del Museo Nacional, México. Xerox of a handwritten paleography of the documents in the handwriting of Paso y Troncoso, loaned by Eva Hunt.
Peñafiel, Antonio, ed.
   1895   *Códice Fernández Leal*. México: Secretaría de Fomento.
Ponce, Padre Alonso
   1967   *Oaxaca en 1568, Descripción Tomada de la Relación Breve y Verdadera de Algunas Cosas que le Sucedieron al Padre Fray Alonso Ponce en las Provincias de la Nueva España*. México: Andrés Henestrosa.
Pesman, Walter
   1962   *Meet Flora Mexicana*. Six Shooter Canyon, Globe, Ariz.: Dale Stuart King.
Poleman, Thomas T.
   1964   *The Papaloapan Project*. Stanford: Stanford University Press.
Price, Barbara
   1971   Prehispanic Irrigation Agriculture in Nuclear America. *Latin American Research Review* 6(3):3-60.
Ramírez Cantú, Débora
   1948   Anotaciones Generales Sobre la Vegetación Acuática, Ruderal, y Arvense de Cuicatlán y sus Alrededores. *Anales del Instituto de Biología de México* 19(2):427-40.
Redfield, Robert
   1971   *The Little Community*. Chicago: University of Chicago Press.
Redmond, Elsa M.
   1983   A Fuego y Sangre: Early Zapotec Imperialism in the Cuicatlán Cañada, Oaxaca. *Studies in Latin American Ethnohistory & Archaeology*, Joyce Marcus, gen. ed., Vol. 1. *Memoirs of the Museum of Anthropology, University of Michigan* 16.
Redmond, Elsa M., and Charles S. Spencer
   1983   The Cuicatlán Cañada and the Period II Frontier of the Zapotec State. In *The Cloud People: Divergent Evolution of the Zapotec and Mixtec Civilizations*, Kent V. Flannery and Joyce Marcus, eds., pp. 117-20. New York: Academic Press.
Relación de Atlatlauca y Malinaltepec
   1905   Relación de Atlatlauca y Malinaltepec...por Francisco de la Mezquita....In *Papeles de Nueva España, Segunda Serie, Geografía y Estadística*, Francisco del Paso y Troncoso, ed., Vol. 4:163-76. Madrid: Est. Tipográfico "Sucesores de Rivadeneyra."
Relación de Cuicatlán
   1905   Relación de Cuicatlán...por Juan Gallego.... In *Papeles de Nueva España, Segunda Serie, Geografía y Estadística*, Francisco del Paso y Troncoso, ed., Vol. 4:183-89. Madrid: Est. Tipográfico "Sucesores de Rivadeneyra."
Relación de Guautla
   1962   Relación de Guautla. *Tlalocan* 4:3-16.
Relación de los Obispados de Tlaxcala, Michoacán, Oaxaca y Otros Lugares....
   1904   *Relación de los Obispados de Tlaxcala, Michoacán, Oaxaca, y Otros Lugares en el Siglo XVI*. Luís García Pimentel, ed. México: Casa de Editor.
Relación de Papaloticpac
   1905   Relación de Papaloticpac y su Partido...por Pedro de Navarrete.... In *Papeles de Nueva España, Segunda Serie, Geografía y Estadística*, Francisco del Paso y Troncoso, ed., Vol. 4:88-99. Madrid: Est. Tipográfico "Sucesores de Rivadeneyra."
Relación de Tepeucila—see Relación de Papaloticpac, pp. 93-99.
Relaciones Geográficas del Siglo XVIII—see Paso y Troncoso, ed., n.d.
Romero, Matías
   1886   *El Estado de Oaxaca*. Barcelona: Tip. Litografía de Espasa y Comp.
Ruíz Oronoz, Manuel, and Teofilo Herrera
   1948   Levaduras, Hongos Macroscópicos, Líquenes y Hepáticas Colectados en Cuicatlán, Oaxaca. *Anales del Instituto de Biología de México* 19(2):299-316.
Russell, J.C.
   1958   Late Ancient and Medieval Population. *Transactions of the American Philosophical Society*, n.s., 48(3).
Sabloff, Jeremy A., and C.C. Lamberg-Karlovsky, eds.
   1974   *The Rise and Fall of Civilizations; Modern Archaeological Approaches to Ancient Culture*. Menlo Park: Cummings Publishing Co.

Sanders, William T.
1956 The Central Mexican Symbiotic Region: A Study of Prehistoric Settlement Patterns. In *Prehistoric Settlement Patterns in the New World*, Gordon R. Willey, ed., pp. 115-27. Viking Publications in Anthropology 23. New York: Wenner-Gren Foundation for Anthropological Research.
1957 Tierra y Agua (Soil and Water): A Study of Ecological Factors in the Development of Meso-American Civilizations. Ph.D. dissertation, Harvard University.
1972 Population, Agricultural History, and Societal Evolution in Mesoamerica. In *Population Growth: Anthropological Implications*, Brian J. Spooner, ed., pp. 101-52. Cambridge: M.I.T. Press.

Sanders, William T., and Barbara J. Price
1968 *Mesoamerica: The Evolution of a Civilization*. New York: Random House.

Sanders, William T., Jeffrey R. Parsons, and Robert S. Santley
1979 *The Basin of Mexico: The Cultural Ecology of a Civilization*. New York: Academic Press.

Secretaría de Agricultura y Fomento
1939 Atlas Climatológico de México. México: Dirección de Geografía, Meteorología, e Hidrología; Servicio Meteorológico Mexicano.

Seler, Eduard
1906 Einige fein bemalte alte Thongefässe de Dr. Sologuren'schen Sammlung aus Nochistlán und Cuicatlán im Staate Oaxaca. *Compte Rendu de la XVème Session du Congrès International des Américanistes, Québec* 2:391-403.

Smith, C. Earle, Jr.
1965a Agriculture, Tehuacán Valley. *Fieldiana, Botany* 31:49-100.
1965b Flora, Tehuacán Valley. *Fieldiana, Botany* 31:101-42.

Spencer, Charles S.
1982 *The Cuicatlán Cañada and Monte Albán. A Study of Primary State Formation*. New York: Academic Press.

Spencer, Charles S., and Elsa M. Redmond
1979 Irrigation, Administration, and Society in Formative Tehuacán. In Prehistoric Social, Political, and Economic Development in the Area of the Tehuacán Valley, Robert D. Drennan ed., pp. 201-15. *Museum of Anthropology University of Michigan, Technical Reports* 11.
1983 A Middle Formative Elite Residence and Associated Structures at La Coyotera, Oaxaca. In *The Cloud People: Divergent Evolution of the Zapotec and Mixtec Civilizations*, Kent V. Flannery and Joyce Marcus, eds., pp. 71-72. New York: Academic Press.

Spooner, Brian, ed.
1972 *Population Growth: Anthropological Implications*. Cambridge: M.I.T. Press.

Spores, Ronald
1967 *The Mixtec Kings and Their People*. Norman: University of Oklahoma Press.
1969 Settlement, Farming Technology, and Environment in the Nochixtlan Valley. *Science* 166:557-69.
1972 An Archaeological Settlement Survey of the Nochixtlan Valley, Oaxaca. *Vanderbilt University Publications in Anthropology* 1.
1983 Postclassic Mixtec Kingdoms: Ethnohistoric and Archaeological Evidence. In *The Cloud People: Divergent Evolution of the Zapotec and Mixtec Civilizations*, Kent V. Flannery and Joyce Marcus, eds., pp. 255-60. New York: Academic Press.

Spores, Ronald, and Miguel Saldaña
1973 Documentos para la Etnohistoria del Estado de Oaxaca: Índice del Ramo de Mercedes del Archivo General de la Nación, México. *Vanderbilt University Publications in Anthropology* 5.

Starr, Frederick
1900 Notes upon the Ethnography of Southern Mexico. *Proceedings of the Davenport Academy of Natural Science* 8, 9.

Steward, Julian H.
1950 Area Research: Theory and Practice. Social Science Research Council Bulletin 63. New York: Social Science Research Council.

Suma de Visitas
1905 Suma de Visitas de Pueblos. In *Papeles de Nueva España, Segunda Serie, Geografía y Estadística*, Francisco del Paso y Troncoso, ed., Vol. 1. Madrid: Est. Tipográfico "Sucesores de Rivadeneyra."

Taylor, William B.
1972 *Landlord and Peasant in Colonial Oaxaca*. Stanford: Stanford University Press.

Theatro Americano
1952 *Theatro Americano, Descripción general de los Reynos de Provincias de la Nueva-España, y sus Jurisdicciones*, (1748) por José Antonio Villaseñor y Sánchez. 2 Vols. México: Editora Nacional, S.A.

Thiéry de Menonville, Nicolas Joseph
1812 Travels to Guaxaca. In *Voyages and Travels in All Parts of the World*, John Pinkerton, ed., Vol. 8. London: Longman, Hurst, Rees, Orme, and Brown.

Torquemada, Juan de
1723 *Los Veinte i un Libros Rituales i Monarchía Indiana....* 3 Vols. Madrid. (Reprinted 1975, México: Editorial Porrúa, S.A.)

Tributos de Pueblos de Indios
1940 Tributos de Pueblos de Indios (Virreinato de Nueva España) 1560. *Boletín del Archivo General de la Nación* 11(1). México.

Villaseñor y Sánchez, José Antonio—see *Theatro Americano*.

Weitlaner, Robert J.
1961 *Datos Diagnósticos para la Etnohistoria del Norte de Oaxaca*. México: Instituto Nacional de Antropología e Historia, Dirección de Investigaciones Antropológicas, Publicación 6.
1969 The Cuicatec. In *Handbook of Middle American Indians*, Robert Wauchope, gen. ed., Vol. 7, Ethnology, Evon Z. Vogt, ed., pp. 434-47. Austin: University of Texas Press.

Winter, Marcus C.
1975 Reportes Breves de Excavaciones. La Cañada. *Boletín* (Centro Regional de Oaxaca, INAH) 2, May 1975.

Wittfogel, Karl A.
1955 Developmental Aspects of Hydraulic Societies. In *Irrigation Civilizations: A Comparative Study*. Julian H. Steward, ed., pp. 43-52. Social Science Monographs 1. Washington, D.C.: Pan American Union.
1957 *Oriental Despotism. A Comparative Study of Total Power*. New Haven: Yale University Press.
1972 The Hydraulic Approach to Pre-Spanish Mesoamerica. In *The Prehistory of the Tehuacán Valley*, Vol. 4: *Chronology and Irrigation*, Freder-

ick Johnson, ed., pp. 59-80. Austin: University of Texas Press.

Wolf, Eric R.
1956 Aspects of Group Relations in a Complex Society: Mexico. *American Anthropologist* 58:1065-78.
1957 Closed Corporate Peasant Communities in Mesoamerica and Central Java. *Southwestern Journal of Anthropology* 13:1-18.

Wolf, Eric R., and Ángel Palerm
1955 Irrigation in the Old Acolhua Domain, Mexico. *Southwestern Journal of Anthropology* 11:265-81.

Woodbury, Richard B., and James A. Neely
1972 Water Control Systems of the Tehuacán Valley. In *The Prehistory of the Tehuacán Valley*, Vol. 4: *Chronology and Irrigation*, Frederick Johnson, ed., pp. 81-153. Austin: University of Texas Press.

Resumen en Español

Capítulo I. Introducción: Cuicatlán, la Cañada, y los Cuicatecos

En este libro se examina un sistema de riego desde tiempos prehispánicos al presente. El sistema se estudia como parte de la región cuicateca. Al norte el límite de la región es la Cañada del Río Santo Domingo. Al este la frontera entre los chinantecos y los cuicatecos queda en la pendiente oriental de la Sierra Madre Oriental. Al sur la frontera entre los zapotecos y los cuicatecos yace a las espaldas del Valle de Oaxaca. Al oeste el fondo de la Cañada fue cuicateco, mientras que los pueblos en la sierra al oeste de la Cañada fueron mixtecos.

La región cuicateca consiste en dos partes: la Cañada, en *tierra caliente*, y la sierra cuicateca, en *tierra fría* y *tierra templada*. La Cañada se caracteriza por vegetación xerofítica como mezquite, palo verde, y el cardón. Es necesario el riego para la agricultura. Con el riego se cultivan maíz, frijoles, jitomates, y muchos palos de frutas, como chicozapotes. Importantes plantas introducidas después de la conquista incluyen el mango y la caña de azúcar. En la sierra, la vegetación natural consiste en robles y pinos. Sólo es posible la agricultura de temporal. Además de maíz, frijoles, y calabazas, se cosechan nueces y un poco de café. Por métodos antiguos se corta madera que se vende en la Cañada en forma de vigas, tablas, y carbón. También la Papelería Tuxtepec está cortando pinos en la sierra para la manufactura de papel.

La Cañada siempre ha servido como rumbo natural de comunicación. Hoy el ferrocarril y una carretera pasan por la Cañada.

Capítulo II. El Ecosistema Cuicateco: Una Reconstrucción Histórica

Al tiempo de la Conquista, los *cacicazgos* cuicatecos participaron de un sistema con tres niveles de integración. El nivel más bajo fue el *cacicazgo*. El nivel más alto fue el macrosistema de Mesoamérica. A un nivel intermedio hubo una integración de los *cacicazgos* cuicatecos entre sí y con *cacicazgos* no cuicatecos en algunos casos.

Los *cacicazgos* de la sierra dependían de agricultura de temporal. Producían algunos productos especializados, como pulque, madera, y productos derivados de la caza. Algunos pueblos tenían recursos únicos como la sal y el oro.

Los *cacicazgos* de la Cañada dependían enteramente de la agricultura de riego. Los pueblos de la Cañada se especializaban en la producción de cosechas de alto valor económico como las frutas, las mismas que comerciaban por toda la región. Además, de los asentamientos de los pueblos de la Cañada estos desarrollaron la vía de comunicación por esta región, como una de las rutas importantes de Mesoamérica.

Las adaptaciones económicas de los pueblos de la sierra y la Cañada formaron la base de las relaciones simbióticas. Estas relaciones se reforzaban por alianzas de parentesco formadas entre *caciques*, quienes a su vez, fueron exógamos de sus propios pueblos. Las alianzas no se limitaban a cuicatecos; pueblos cuicatecos muchas veces se ligaron con pueblos fuera de los límites del sistema cuicateco.

Finalmente, el sistema cuicateco formó parte del imperio azteca al tiempo de la Conquista. Los efectos del macrosistema se manifestaban en los niveles de tributo y condiciones de intercambio.

Capítulo III. Los Cuicatecos en Contacto Cultural

Se analiza el efecto del contacto entre la cultura española y la cultura cuicateca, siguiendo el modelo presentado en "Culture Contact and Schismogenesis," por Gregory Bateson (1972). Se pueden caracterizar todas las relaciones entre simétricas y complementarias. El efecto de la Conquista trató de establecer la simetría de las culturas del patrón "indio." La estructura de las comunidades indígenas se caracterizaba por simetría, mientras las relaciones entre los indios y los españoles eran complementarias. Los pueblos españoles, en contraste, tenían estructuras complementarias.

El efecto de despoblación y la introducción de nuevos elementos por los españoles, seguían la estructura del ecosistema. Se estudia el contacto entre pueblos simétricos del patrón "indio" en contraste con los pueblos complementarios del patrón "español." En las primeras décadas, la Cañada y los pueblos cuicatecos atrajeron pocos españoles. Así el sistema cuicateco quedaba simétrico e "indio." La aceptación de elementos introducidos por los españoles dependía en su compatibilidad con un sistema simétrico. De esta manera, el ganado menor era compatible con una vida simétrica, especialmente con la bajada de población, mientras que el ganado mayor se asociaba con un sistema complementario español.

Capítulo IV. El Ecosistema Cuicateco desde la Conquista a la Actualidad

En los cuatrocientos años desde la Conquista podemos observar la interacción de factores en los procesos que afectan el ecosistema cuicateco.

Siglo XVII: Introducción de la caña de azúcar. En el siglo XVII la población del ecosistema llegó a su punto más bajo. Aparecieron las primeras referencias del uso de trapiches. La producción de azúcar demandaba una estructura complementaria para su explotación. Muchas veces se importaron trabajadores negros para el cultivo de la caña. La caña funcionaba como fuerza económica en la dirección del patrón complementario: "español."

Siglo XVIII: Los españoles empiezan a establecerse en la Cañada. Dos factores favorecían el establecimiento de españoles en la Cañada durante el siglo XVIII: primero, el azúcar, que no era compatible con el patrón simétrico de los indios; y el segundo, la presencia del comercio de grana y otros productos que circulaban por los pueblos en el Camino Real de la Cañada. Así, algunos españoles se establecieron en la Cañada para tratar con los productores de los pueblos de la sierra. De tal manera, tanto el Camino Real y como la caña de azúcar tenían la razón de favorecer el patrón complementario y "Español" en la Cañada, sin afectar mucho a los pueblos de la sierra.

Siglo XIX: Caminos, ferrocarriles, y haciendas. Hasta el siglo XIX el Camino Real por la Cañada fue transitable solamente a pie o a caballo. Empezando en 1833, y terminando con los esfuerzos de Juárez desde 1850 a 1865, fue mejorado el Camino Real para permitir el pasaje de vehículos. En 1892 se construyó el primer ferrocarril que cruza la Cañada. La consecuencia del mejoramiento de las vías de transporte en la Cañada fue la de exagerar las tendencias complementarias en los pueblos de esta región. En los pueblos del patrón "Indio" de la Cañada, se había añadido un estrato formado por los comerciantes, agentes del gobierno, y otros españoles quienes trataban a veces con los pueblos de la sierra, pero vivían en la Cañada y se relacionaban con el macrosistema por medio del Camino Real. Al fin del siglo XIX habían dos tipos de pueblos en la Cañada: las haciendas, y los pueblos. Estos segundos están formados por dos estratos sociales de habitantes: un estrato bajo indígena; y un estrato alto de comerciantes, oficiales, y otros quienes trataban con los pueblos de la sierra desde el Camino Real. En tal sentido, solo quedaron pueblos complementarios de "Españoles" en la Cañada, mientras los pueblos simétricos seguían en la sierra.

El ecosistema cuicateco en el siglo XX. Hoy en día el ecosistema Cuicateco constituye el resultado de los procesos económico-sociales que hemos observado desde la Conquista. Los pueblos de la Cañada corresponden al patrón "Español," caracterizados por una estructura complementaria. Su economía está basada, completamente en la agricultura de riego. Los pueblos que eran haciendas todavía dependen mucho de la caña de azúcar. Los otros pueblos de la Cañada concentran mayormente su terreno en cultivos de fruta y legumbres de valor comercial en los mercados fuera de la Cañada. La población de hoy se acerca a la misma estimada para el tiempo anterior de la Conquista. El patrón de agricultura, con producción de cosechas especiales por cada región amplia, también se asemeja al período anterior a la Conquista.

Los pueblos de la sierra quedaron siempre como simétricos, con el patrón "indio." Sus cosechas son básicamente para el uso de los mismos pueblos. Nueces y café, producidos en la sierra, se venden a comerciantes de los pueblos de la Cañada. Madera es tallada en la sierra; y luego es vendida en la Cañada; o es explotada por una compañía moderna, como la Papelería Tuxtepec, con oficinas en Cuicatlán.

Sólo los pueblos en la Cañada tienen un "elite." Los individuos y familias ricos que controlan el poder en la Cañada tienen mucha influencia en el manejo de sistemas de riego. Sin embargo, parece que su poder y riqueza vienen, no tanto del control de riego, o de terrenos de riego, sino del control del comercio, pasando de la sierra por intermedio de los pueblos de la Cañada hacia los mercados regionales. Otro factor importante en los pueblos de la Cañada es la presencia de centros de población grande, accesibles mediante una red de transporte que sirve principalmente a los mercados de frutas y legumbres producidas mediante la agricultura de riego en la Cañada. Hoy, como antes, los pueblos de la Cañada existen, no solamente para sí, sino también para toda la región.

Capitulo V. Evidencia Arqueológica: El Ecosistema Cuicateco desde su Origen

La arqueología de la Cañada se ha dividido en cuatro períodos: fase Perdido (del formativo medio), fase Lomas (formativo tardío y terminal), fase Trujano (clásico), y fase Iglesia Vieja (postclásico al contacto). La evidencia arqueológica sugiere que los primeros asentamientos se basaron en el riego aprovechando del aluvión bajo a la orilla del Río Grande. En la fase Lomas, con la intervención directa de los zapotecos, se forzó el restablecimiento de los pueblos al pie del monte y se construyeron un sistema de riego utilizando el aluvión alto. Un canal en el sitio de La Coyotera se ha fechado en la fase Lomas, y hay evidencia de la intervención del Estado zapoteco en forma de un *tzompantli*. En la siguiente fase Trujano, los zapotecos renunciaron el control directo de la Cañada, pero la población del lugar seguía usando el aluvión alto. Poco a poco la población crecía. Correspondiente al postclásico no se encontró ningún sitio en terreno aprovechable por el riego, aunque en muchos lugares se encontraron canales más altos que los que están en uso hoy en día, implicando sistemas más extensivos que los actuales. Inclusive se localizó un complejo, de testimonios en El Despoblado, que consiste en un canal pegado a un escarpado llevando agua a una serie de terrazas artificiales por toda esta área.

El análisis de los pueblos basados en riego en la Cañada sugiere que la mayor importancia fue alcanzado durante el postclásico, cuando formaron parte del proceso de crecimiento de la población.

El análisis de la relación entre riego y estratificación social no apoya el papel causal del riego. El riego en la Cañada tiene por causa la intervención de una sociedad estratificada. Estas fueron los zapotecos. En la Cañada la estructura estratificada se relacionaba siempre con el control de intercambio entre el macrosistema y los pueblos de la región cuicateca. Los pueblos en la Cañada inevitablemente tenían que ser el foco de intercambio, por cuanto quedaban en la ruta de intercambio, por estar ligado por el Camino Real. Por esta razón, el poder venía, no del canal de riego directamente, sino del control de comercio y tributo, tanto ayer como hoy.

Apéndice I. Trabajo Realizado en el Campo

El trabajo de campo consistía en recorridos extensivos de superficie alrededor de Cuicatlán, una exploración de diez días pasando por los otros pueblos de la Cañada, y dos excavaciones chicas en Cuicatlán.

Los reconocimientos de superficie cerca de Cuicatlán dieron por resultado la identificación de 31 sitios, un mapa de cada uno de ellos, y la colección y análisis de artefactos de cada yacimiento para su fechamiento. Casi todos los sitios pertenecían a la fase Iglesia Vieja. Sólo uno, el sitio 11, pertenecía a la fase Perdido. Este lugar se encontró en la primera terraza arriba del Río Grande. Los demás sitios se hallaron en terrenos arriba de los canales de riego, para no usar los terrenos que bien fueron utilizados en la agricultura de riego.

Se efectuaron dos excavaciones. La primera se hizo en el sitio 5, donde restos de un canal cortaba un lado de cimientos de una casa. Se encontraron restos de una casa con un piso de yeso. Bajo el piso se hallaron dos tumbas. Las paredes de la casa tenían la misma orientación que el canal, indicando contemporaneidad, en la fase Iglesia Vieja. Excavaciones en un segundo sitio, identificado con el número 16, tenía como meta buscar una columna estratificada más profunda, y dentro de ellas restos de vegetales colectados mediante flotación. La excavación en el sitio permitió encontrar evidencia de dos casas superpuestas, ambas con pisos de yeso como en el ejemplo anterior. Ambas casas pertenecían a la fase Iglesia Vieja.

Luego, se hizo un reconocimiento de diez días de toda la Cañada, guiado por Rafael Vásquez Cruz, inspector arqueológico del I.N.A.H. para esta área. Con su ayuda, se hallaron restos de muchos asentamientos y canales postclásicos. Todos los casos muestran un patrón de conservar los terrenos aprovechables por el riego, y más bien utilizar para las edificaciones de viviendas, suelos por encima de ellas. Dos sitios que localizamos pertenecen a períodos más tempranos. El sitio de La Coyotera, que fue investigado por Charles Spencer (1982), pertenece a las fases Perdido y Lomas. El sitio de La Unión, cerca a Atlatlauca, pertenece a la fase Lomas.

Toda el material coleccionado se depositó en el Museo de Antropología de Oaxaca.